Rosita Sowade

Continuous-wave terahertz light from optical parametric oscillators

Rosita Sowade

Continuous-wave terahertz light from optical parametric oscillators

Südwestdeutscher Verlag für Hochschulschriften

Imprint
Any brand names and product names mentioned in this book are subject to trademark, brand or patent protection and are trademarks or registered trademarks of their respective holders. The use of brand names, product names, common names, trade names, product descriptions etc. even without a particular marking in this work is in no way to be construed to mean that such names may be regarded as unrestricted in respect of trademark and brand protection legislation and could thus be used by anyone.

Publisher:
Südwestdeutscher Verlag für Hochschulschriften
is a trademark of
Dodo Books Indian Ocean Ltd., member of the OmniScriptum S.R.L Publishing group
str. A.Russo 15, of. 61, Chisinau-2068, Republic of Moldova Europe
Printed at: see last page
ISBN: 978-3-8381-2695-1

Zugl. / Approved by: Bonn, Rheinische Friedrich-Wilhelms-Universität Bonn, Dissertation, 2010

Copyright © Rosita Sowade
Copyright © 2011 Dodo Books Indian Ocean Ltd., member of the OmniScriptum S.R.L Publishing group

Contents

1 Introduction   1

2 Fundamentals   5
    2.1 Optical parametric amplification 6
    2.2 Optical parametric oscillation 13
    2.3 Quasi phase matching 18
    2.4 Terahertz wave generation with cascaded parametric processes 22

3 Output power optimisation   25
    3.1 Experimental methods for the infrared setup 25
    3.2 Efficiency characterisation of the parametric oscillator 30
    3.3 Discussion of performance optimisation 34

4 Spectral features of the resonant waves   39
    4.1 Primary parametric process 39
    4.2 Spectral characteristics for processes involving terahertz waves 42

5 Terahertz wave generation   49
    5.1 Experimental methods for the terahertz setup 49
    5.2 Terahertz optical parametric oscillator 55
    5.3 Discussion of terahertz wave characterisation 67
    5.4 Material properties of lithium niobate in the infrared and terahertz frequency regime 77
    5.5 Comparison of system performance with that of other methods 78

6 Summary	**81**
Bibliography	**83**

CHAPTER 1

INTRODUCTION

One of the current gaps in physics requiring a bridge to conceal it can be found in the electromagnetic spectrum. In 1888 Heinrich Hertz showed the equivalence of electromagnetic waves and light waves. Since then most frequency regions have been made accessible by technologies, providing sources as well as detectors for certain wavelengths, but the so-called *terahertz gap* remained untouched for a long time [1]. This terahertz range can be defined as frequencies between 0.1 and 10 THz [2], while one terahertz comprises 10^{12} oscillations per second, lying thus between microwaves (below 0.1 THz) and the infrared (above 10 THz).

Why would a frequency region, which is surrounded by frequencies that are widely used in devices such as radios, mobile phones or remote controllers, be left unexplored? Certainly not because of a lack of interesting phenomena. Whereas rotations and oscillations of single molecules can be found in the near and mid infrared [3], interactions between molecules generate radiation in the terahertz range [4,5]. For example, half of the light, being sent towards the earth from the milky way, is thus in the terahertz regime containing valuable information for astronomy [6]. Terahertz spectroscopy is also very useful for chemical analysis [7–9]. Recently, terahertz photonics also became of interest for applications in biology and medicine [10], security technologies [11] and quality control, for example of polymeric products [12].

Two conceptionally different types of coherent light sources, and therefore also of coherent terahertz radiation, can be distinguished: pulsed and continuous-wave (cw). Due to high peak intensities in light pulses, most technologies are explored first with pulsed systems. Thus various pulsed terahertz sources already exist [13,14]. For certain applications such as astronomy [15] or communications [16–18], however, continuous-wave operation is desirable because of its small linewidth and continuous carrier wave to serve as local oscillators or for information transfer. Hence, this thesis concentrates entirely on the generation of continuous-wave terahertz light. Systems, based on different physical concepts, have recently been developed in this area. Some approaches start from electronic frequencies whereas others rely on optical methods.

INTRODUCTION

Beginning with low frequencies, electronic multipliers and backward wave oscillators can be employed to reach the terahertz regime [19–21]. Yet, these devices will not be able to span the entire terahertz range, because their achievable frequencies are restricted by carrier lifetimes, leading to a strong frequency roll-off towards higher THz frequencies, and so far no frequencies above 3 THz have been reached [22]. Additionally, the tuning of a single source usually amounts to only 20 % around the center frequency [23]. Combinations of optics and electronics – opto-electronic systems – have similar frequency constraints [24]. Nevertheless, such devices are currently widely used. They base on photoconductive antennas, so-called photomixers, which are excited by two laser beams with different wavelengths generating a wave of their difference frequency [25]. Here, the tuning is restricted by the tunability of the two lasers.

Taking a look at optical terahertz systems: direct optical lasers exist for the terahertz range, being able to provide high powers up to watts with frequencies between 0.3 to 10 THz, but such devices emit discrete lines and are usually not tunable at all [23, 26]. For most applications, however, terahertz output powers in the order of micro- or milliwatts are sufficient. In 2002, terahertz quantum cascade lasers, relying on semiconductor structures, were developed [27] but their beam profile characteristics are challenging to optimise [28, 29]. In addition, they need cryogenic temperatures for operation and can hardly produce radiation with frequencies below 1 THz [30].

The field of *nonlinear optics* comprises multiple advantages for frequency conversion to any desired wavelength. It does not contain inherent frequency boundaries and can provide widely tunable sources. One approach for terahertz wave production with nonlinear optics is optical difference frequency generation. Here, two lasers are sent through a nonlinear medium which creates the difference frequency of the two wavelengths. Thus, tunability and achievable frequencies are limited by the tuning of the two lasers and so far only some nanowatts of output power could be generated [31, 32].

Optical parametric oscillators (OPOs) are more versatile and known for their wide tuning ranges [33, 34]. They require only one pump laser, whose light is converted within a nonlinear material into signal and idler wave. The resonance condition requires that the frequency of the pump wave at ν_p and the sum of signal and idler frequencies, $\nu_\mathrm{s} + \nu_\mathrm{i}$, are identical [35]. Recently, singly-resonant systems, in which only the signal wave oscillates, based on periodically-poled lithium niobate have become working horses for spectroscopy and high-power applications [36, 37]. Although these devices are now even commercially available, matters of improvement and open questions remain.

The overall goal of this work is to extend the frequency range of OPOs to the THz regime. Hence, the thesis is organised as follows: Chapter 2 gives an introduction to the theoretical concepts of optical parametric oscillation in general, quantifying solutions of the coupled wave equations including absorption. Additionally, phase matching schemes for terahertz generation are explained. The next chapter elucidates the influence of the crystal length on maximum achievable output powers in standard optical parametric oscillators. High idler output powers also correspond to high signal powers which are needed for terahertz generation due to the large absorption of THz waves in nonlinear crystals. Chapter 4 deals with the clarification of spectral features in the resonant waves in singly-resonant OPOs at high pump powers. The considerations of all these insights form the basis for the realisation of the first cw terahertz optical parametric oscillator. In Chapt. 5, its performance in terms of terahertz output power, beam profile and tuning properties is characterised. Furthermore, the material properties of lithium niobate such as the nonlinear coefficient and temperature dependence of the refractive index in the terahertz regime are analysed. The significance of this terahertz source in comparison with that of other methods of THz generation is discussed.

Chapter 2

Fundamentals

The major scope of this work is the generation of monochromatic terahertz light with the means of nonlinear optics. The efficiency of such a frequency conversion process is higher with larger nonlinearities of the material, but decreases with growing losses. Therefore, nonlinear optics is usually performed only in the transparency range of a medium where losses due to absorption can be neglected. However, resonances of lattice vibrations in nonlinear crystals are in the terahertz frequency range [38], causing strong losses.

Standard OPO theories neglect absorption of the interacting waves [39]. Some efforts to extend the theory have been performed [40], but in this thesis calculations are presented that are more general in some features while being tailored to our problems in other respects. First, the concept of optical parametric amplification is introduced, providing solutions of the coupled wave equations including absorption. These consideration are then extended to optical parametric oscillation, giving measures of the efficiency of such a process. Afterwards ways of calculating the oscillation threshold with and without absorption are demonstrated.

For efficient frequency conversion the resonance and the phase matching conditions need to be satisfied while the latter one is not naturally fulfilled in dispersive media. Therefore, quasi phase matching is explained, illustrating the processes relevant for our devices. In particular, a backwards parametric process is introduced. Example calculations are performed for the nonlinear material lithium niobate, since this is the one used in this thesis.

All these aspects contribute to the realisation and analysis of a terahertz optical parametric oscillator whose fundamental concept is presented in the final section of this chapter, relying on a cascaded parametric process.

FUNDAMENTALS

2.1 Optical parametric amplification

A dielectric medium is an electrical insulator with a polarisation $\mathcal{P}$, describing the material's reaction to incident electromagnetic waves with the electric field E. This polarisation can be expressed by the optical susceptibility tensor χ [39]:

$$\mathcal{P}_m(E) = \epsilon_0 \chi^{(1)}_{mn} E_n + \epsilon_0 \chi^{(2)}_{mno} E_n E_o \ . \tag{2.1}$$

Here, ϵ_0 is the vacuum permittivity constant. All higher-order terms are neglected, since these are of no relevance for parametric processes. Only dielectric materials with a non-vanishing contribution of $\chi^{(2)}$ can support such processes. Lithium niobate, the nonlinear medium used in this thesis, has a spontaneous ferroelectric polarisation along one crystallographic axis [41], which is responsible for its second order nonlinearity.

One process based on this second order nonlinearity $\chi^{(2)}$ is *optical parametric amplification* (abbreviated OPA). It is described by a pump wave at a frequency of ν_p and a signal wave at ν_s, generating an idler wave at ν_i (see Fig. 2.1a), with the resonance condition

$$\nu_p = \nu_s + \nu_i \ . \tag{2.2}$$

In the frame of this work, the propagation direction of all interacting waves is reduced to one dimension, z. Both, signal E_s and idler fields E_i, will be amplified throughout the nonlinear medium, but E_i is not present at the front facet of the crystal (see Fig. 2.1b). Beginning and ending of the nonlinear medium are denoted by $z = 0$ and $z = L$, respectively, which makes L the crystal length.

The slope of the rising field in Fig. 2.1b is plotted for non-decreasing pump power, i.e. $E_p(z) = E_p(0) = $ constant. This would cause infinite rise of the generated wave, which of course is not true for the experiment. Calculations including pump depletion will be performed in the section on parametric oscillation.

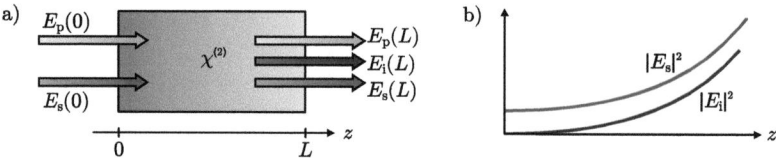

Figure 2.1: *a) Scheme of optical parametric amplification. A pump field E_p and a signal field E_s create an idler field E_i within a nonlinear crystal of the length L. b) An incoming signal field E_s is amplified and an idler field E_i is generated at $z = 0$, which grows with the propagation direction z. The pump field E_p is assumed to remain constant over the entire length z.*

2.1.1 Coupled wave equations

The coupled wave equations for OPA, including linear absorption, can be derived from Maxwell's equations [42] and are then given by

$$\frac{\partial}{\partial z}E_\text{p}(z) = -i\gamma_\text{p} E_\text{s}(z) E_\text{i}(z) e^{+i\Delta kz} - \frac{\alpha_\text{p}}{2} E_\text{p}(z) ,$$
$$\frac{\partial}{\partial z}E_\text{s}(z) = -i\gamma_\text{s} E_\text{p}(z) E_\text{i}^*(z) e^{-i\Delta kz} - \frac{\alpha_\text{s}}{2} E_\text{s}(z) ,$$
$$\frac{\partial}{\partial z}E_\text{i}(z) = -i\gamma_\text{i} E_\text{p}(z) E_\text{s}^*(z) e^{-i\Delta kz} - \frac{\alpha_\text{i}}{2} E_\text{i}(z) . \qquad (2.3)$$

The gain coefficient γ_l is defined by $\gamma_l = 2\pi\nu_l d/n_l c$ and contributes to the conversion part of the equations with n_l being the refractive index, d the nonlinear coefficient and c the vacuum velocity of light. Depending on the relative phases this can lead to amplification of some fields due to energy transfer from one wave to another. The α_l-contribution symbolises a loss mechanism due to intensity absorption with l being either the index p of the pump, the signal s or the idler wave i. This linear absorption α_l can be derived from the imaginary part of the linear susceptibility $\chi^{(1)}$ of a medium [43].

The nonlinear coefficient d is proportional to the second order nonlinearity by $\chi^{(2)} = 2d$. In general, these two quantities are third rank tensors as indicated in Eq. (2.1), but in this work we only use the strongest nonlinear contribution of lithium niobate d_{333} where all polarisations of the three interacting waves are extra-ordinary and we thus can set $d_{333} = d$ for simplicity.

In 1964 Miller empirically realised that the ratio between first $\chi^{(1)}$ and second order nonlinearities $\chi^{(2)}$ is nearly constant for non-centrosymmetric crystals, and accordingly an approximative rule for the nonlinear coefficient d was developed [44, 45]:

$$d(\nu_\text{p}, \nu_\text{s}, \nu_\text{i}) = d_0\, \chi^{(1)}(\nu_\text{p})\, \chi^{(1)}(\nu_\text{s})\, \chi^{(1)}(\nu_\text{i}) . \qquad (2.4)$$

Here, d_0 is a constant which is in principle specific for each material, but all dielectrics have very similar values. With this rule, nonlinear coefficients can be estimated simply by knowing the refractive indices for the light frequencies involved, since n can be related to the linear susceptibility via $\text{Re}(\chi^{(1)}) = n^2 + 1$ [43]. Equation (2.4) suggests that materials with larger refractive indices also provide higher nonlinear coefficients. Therefore, we use higher values for d in calculations if terahertz waves are involved in the processes, because there the refractive indices of lithium niobate are more than a factor of two higher than those of the infrared [46, 47]: $d_\text{IR} = 17$ and $d_\text{THz} = 100$ pm/V.

Generally, the phase mismatch Δk is a vector, but reduces to a scalar equation for the constraint of only z-direction propagation of all interacting waves:

$$\Delta k = k_\text{p} - k_\text{s} - k_\text{i} . \qquad (2.5)$$

FUNDAMENTALS

This phase mismatch results from dispersion considering the wave vectors with the absolute value $k_l = 2\pi n_l \nu_l / c$ including the refractive indices n_l. All refractive indices in this work are deduced from the Sellmeier equations for magnesium-doped lithium niobate obtained by Gayer et al. [46] for infrared waves and by Palfalvi et al. [47] for terahertz waves.

To solve the set of Eqs. (2.3), we can first of all assume an undepleted pump wave $E_p(z) = E_p(0) = E_p$. Such an assumption is valid for small pump absorptions and small amplifications (typical values are around 1 %, see also Sec. 2.3.1) and reduces the problem to two coupled equations, which can be rewritten as

$$\left(\frac{\partial}{\partial z} - i\frac{\Delta k}{2} + \frac{\alpha_s}{2}\right) \widetilde{E}_s(z) = -i\gamma_s E_p \widetilde{E}_i^*(z) , \tag{2.6}$$

$$\left(\frac{\partial}{\partial z} + i\frac{\Delta k}{2} + \frac{\alpha_s}{2}\right) \widetilde{E}_i^*(z) = +i\gamma_s E_p^* \widetilde{E}_s(z) \tag{2.7}$$

with $\widetilde{E}_{s,i} = E_{s,i} e^{i\Delta k z/2}$.

The exponential ansatz $\widetilde{E}_{s,i} = c_{s,i} e^{\Gamma z/2}$, corresponding to a plane wave assumption, can be inserted into Eqs. (2.6) and (2.7), leading to

$$\begin{pmatrix} \frac{\Gamma}{2} - i\frac{\Delta k}{2} + \frac{\alpha_s}{2} & i\gamma_s E_p \\ -i\gamma_i E_p & \frac{\Gamma}{2} + i\frac{\Delta k}{2} + \frac{\alpha_i}{2} \end{pmatrix} \begin{pmatrix} \widetilde{E}_s \\ \widetilde{E}_i^* \end{pmatrix} = 0 \tag{2.8}$$

which provides two linearly independent solutions for Γ

$$\Gamma_\pm = -\alpha_+ \pm \left[\alpha_-^2 + i\Delta k(2\alpha_- + i\Delta k) + 4\gamma_s\gamma_i |E_p|^2\right]^{1/2} \equiv -\alpha_+ \pm S . \tag{2.9}$$

Here, the abbreviations $\alpha_\pm = (\alpha_i \pm \alpha_s)/2$ are employed, and the solutions of the coupled wave equations take the form

$$\begin{aligned} \widetilde{E}_s(z) &= c_s^+ e^{(\Gamma_+ z/2)} + c_s^- e^{(\Gamma_- z/2)} , \\ \widetilde{E}_i^*(z) &= c_i^+ e^{(\Gamma_+ z/2)} + c_i^- e^{(\Gamma_- z/2)} . \end{aligned} \tag{2.10}$$

It is still necessary to determine the amplitude coefficients $c_{s,i}^\pm$. Initially, only pump and signal wave are sent into the crystal, making the idler field $E_i(0) = 0$ (see Fig. 2.1). This can be used as one boundary condition $\widetilde{E}_i(0)^* = 0$ together with $\widetilde{E}_s(0) = E_s^0$, providing

$$\begin{aligned} c_s^+ + c_s^- &= E_s^0 , \\ c_i^+ + c_i^- &= 0 . \end{aligned}$$

In addition, Eq. (2.6) can be rearranged, such that it provides an expression for $\widetilde{E}_i^*$ which can be inserted into Eqs. (2.10), resulting in

$$(S - \alpha_- - i\Delta k)c_s^+ + (S + \alpha_- + i\Delta k)c_s^- = 0 .$$

These calculations lead to

$$c_s^+ = \frac{1}{2}E_s^0\left(1 + \frac{\alpha_- + i\Delta k}{S}\right),$$
$$c_s^- = \frac{1}{2}E_s^0\left(1 - \frac{\alpha_- + i\Delta k}{S}\right) \qquad (2.11)$$

for the signal wave coefficients. A similar result can be obtained for $c_i^\pm$ as it will be shown in Sec. 2.1.3.

2.1.2 Parametric gain

From the solutions derived above, the parametric gain G for the signal wave can be determined. This gain is defined by the difference between signal intensity $I_s(L)$ leaving the crystal and the intensity in front of it with respect to the initial signal intensity $I_s(0)$. The signal intensity is proportional to the square of the field $I_s \sim |E_s|^2$. Therefore, the intensity gain is given by

$$\begin{aligned}G &\equiv \frac{I_s(L)}{I_s(0)} - 1 \\ &= \frac{1}{4}\left|\left(1 + \frac{\alpha_- + i\Delta k}{S}\right)e^{\Gamma_+ L/2} + \left(1 + \frac{\alpha_- + i\Delta k}{S}\right)e^{\Gamma_- L/2}\right| - 1.\end{aligned} \qquad (2.12)$$

It should be noted that this parametric gain depends on the initial pump power which is proportional to $|E_p|^2$, the nonlinear coefficient d, the phase mismatch Δk and the absorption α, since all of these parameters are part of S. With growing pump power, crystal length L and nonlinear coefficient, also the gain G will increase, whereas higher absorption causes G to decrease. The maximum pump power is usually fixed by the pump laser available while the absorption and the nonlinear coefficient are determined by the material.

To illustrate the general shape of such a gain profile, we first of all make simplifying assumptions by neglecting the absorption, i.e. $\alpha_i = \alpha_s = 0$. Figure 2.2a shows the behaviour of the signal gain with respect to the product of the phase mismatch Δk and the crystal length L for a process of 1030 nm and 1500 nm light producing 3290 nm radiation. The wavelength λ is related to the light frequency via $\lambda = c/\nu$. The maximum efficiency of this parametric process is reached for perfect phase matching $\Delta k = 0$ and gives a value of 1.53 % for $L = 5$ cm. Here, a pump power of 1 W, a beam radius of 100 µm and a nonlinear coefficient of $d_{IR} = 17$ pm/V are used for the calculation. In all examples of this chapter, 1030 nm will be used as a pump wavelength, since this will be the wavelength of the laser employed in the experiments.

The first two zero points for neglected absorption of the gain profile are at $|\Delta k L| = 2\pi$. This is sometimes defined as the bandwidth of the parametric process. However, if idler absorption is taken into account, the side maxima vanish (see Fig. 2.2a). Therefore, we use the full-width-half-maximum (FWHM) of the gain curve as its frequency bandwidth $\Delta\nu$.

Non-vanishing idler absorption α_i causes the parametric signal gain G to broaden, implying a larger acceptance bandwidth for the parametric process, and its maximum value decreases. For an absorption of $\alpha_i = 1$ cm^{-1} the signal gain at $\Delta k = 0$ is only 0.7 % and thus decreased by a factor of two. This nicely illustrates the coupling of the waves: although only the idler wave is absorbed, the gain of the signal wave is diminished (see Fig. 2.2a).

A different behaviour can be observed if one includes signal absorption and assumes the idler absorption to vanish. Figure 2.2b displays the signal gain profile for $\alpha_s = 10^{-3}$ cm^{-1}. This absorption is four orders of magnitude lower than $\alpha_i = 1$ cm^{-1} but already decreases the maximum gain value by a factor of 15.

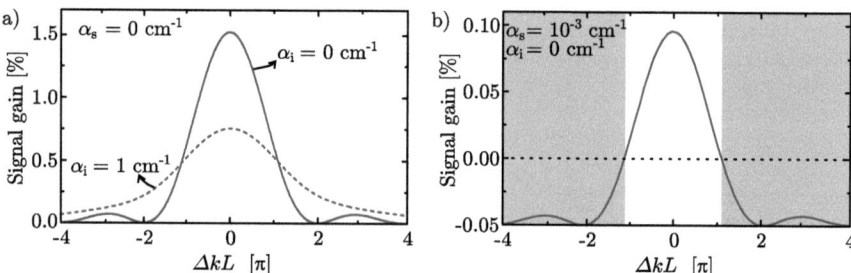

Figure 2.2: *Different influences of signal and idler absorption as seen in the parametric gain G of the signal wave with respect to the phase mismatch Δk multiplied by the crystal length L for a process of 1030 nm and 1500 nm light converted into 3290 nm. a) The signal gain is once plotted for negligible idler absorption (solid line) and once for $\alpha_i = 1$ cm^{-1} (dashed line). In both cases the signal absorption is assumed to vanish. b) Signal gain without idler absorption but a signal absorption of $\alpha_s = 10^{-3}$ cm^{-1}. The grey regions mark regimes where gain turns into loss because it drops below zero.*

Additionally one can see, that part of the gain drops below zero while the profile keeps its overall shape. This implies that high signal absorption will turn the gain into a loss mechanism, even reducing the signal power present at the front facet of the crystal. Therefore, signal absorption should be avoided by all means while idler absorption can be tolerated to a certain extend.

2.1.3 Idler output power including absorption

In the chapter above, we have only considered the amplification of the signal wave within the nonlinear medium. Now, the issue of absorption in connection to the idler output power shall be addressed. Here, signal absorption is assumed to vanish.

To obtain an equation for the idler output, one can plug Eqs. (2.10) into Eq. (2.7), leading to

$$\left(\frac{\Gamma_+}{2} + i\frac{\Delta k}{2} + \frac{\alpha_i}{2}\right) c_i^+ + \left(\frac{\Gamma_-}{2} + i\frac{\Delta k}{2} + \frac{\alpha_i}{2}\right) c_i^- = i\gamma_i E_p^* \widetilde{E}_s^0 . \tag{2.13}$$

Analogously to Eq. (2.11) for the signal coefficients, the coefficients c_i can be determined from the boundary conditions

$$\begin{aligned} c_i^+ &= +i\frac{\gamma_i E_p^* E_s^0}{S}, \\ c_i^- &= -i\frac{\gamma_i E_p^* E_s^0}{S} . \end{aligned} \tag{2.14}$$

This results in an idler output power of

$$P_i = \frac{2n_i}{n_p n_s c\epsilon_0 A} \frac{\gamma_i^2 P_p P_s^0}{|\alpha_-^2 + i\Delta k(2\alpha_- + i\Delta k) + 2\gamma_i\gamma_s|E_p|^2|} \left| e^{\Gamma_+ L/2} - e^{\Gamma_- L/2} \right|^2 . \tag{2.15}$$

Here, it is postulated that all interacting waves are plane waves, with homogeneous intensity distributions and the same beam cross-section A, to convert fields E and intensities $I = cn\epsilon_0 |E|^2/2$ into powers $P/A = I$.

In principle, all waves experience losses due to reflections at the boundary between crystal and air. However, our crystals are anti-reflection coated and therefore we assume all transmission coefficients $T_l = 1$ for the infrared waves. If the output wave is in the terahertz frequency regime this coefficient would have to be included as an additional factor in Eq. (2.15):

$$T_{\text{THz}} = \frac{4 n_{\text{THz}}}{(n_{\text{THz}} + 1)^2} . \tag{2.16}$$

For a refractive index of $n_{\text{THz}} = 5$ the transmission through the exit face of the crystal is only $T_{\text{THz}} = 55\ \%$.

To illustrate the behaviour of the idler output from a nonlinear crystal such as lithium niobate, we plot P_i for different idler absorptions α_i and perfect phase matching $\Delta k = 0$. The signal absorption (see Fig. 2.3) is neglected, thus reducing α_- to α_i in the equations. Exemplary, a process generating 1.5 THz from 1030 and 1035 nm is chosen while the nonlinear coefficient is taken to be $d_{\text{THz}} = 100$ pm/V. The incident powers of signal and pump wave are 1 W.

F<small>UNDAMENTALS</small>

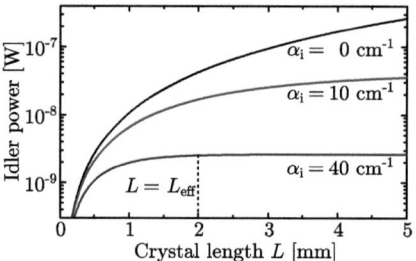

Figure 2.3: *Idler output power for different absorption values α_i, plotted with respect to the crystal length. Larger absorption decreases the resulting power. For high idler absorptions such as $\alpha_i = 40$ cm^{-1}, the idler power saturates at an effective crystal length $L = L_{\text{eff}}$ below 5 mm. Details of the calculation can be found in the text.*

One can see, that the idler power for $\alpha_i = 40$ cm^{-1} is lowered by two orders of magnitude and saturates after a certain crystal length (see Fig. 2.3, right side). This is due to a balance of the parametrically generated and the linearly absorbed idler wave. Thus, for processes with high absorption, there exists an effective crystal length, above which no increase in output power can be achieved. This effective length decreases with growing absorption. Such an effect suggests that one can choose shorter crystals, since above L_{eff} no increase of the terahertz power occurs. The issue of crystal length in connection with output power is also discussed in Chapt. 3 from a different point of view.

2.2 Optical parametric oscillation

Optical parametric oscillation bases on parametric amplification and the same coupled wave equations (see Eqs. (2.3)), but here no initial signal wave has to be sent into the crystal. Both, signal and idler fields, build up from noise [48] and after being amplified through the crystal, the signal wave is coupled back to $z = 0$, as displayed in Fig. 2.4. This feedback can be realised with a resonator, consisting of mirrors. However, since the cavity will not be completely ideal, the resonant intensity experiences losses V and thus the signal field E_s has losses of $\sqrt{V}$. The gain within the crystal must be larger than the losses on one roundtrip for an enhancement to occur. Consequently, the signal power can grow larger than the incoming pump power within the cavity.

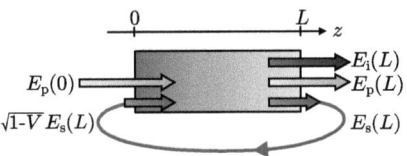

Figure 2.4: *Scheme of optical parametric oscillation. An incoming pump field E_p creates a signal E_s and an idler field E_i in a nonlinear crystal of length L. The signal field is feeded back and experiences losses $\sqrt{V}$.*

Such a device is called singly-resonant *optical parametric oscillator*, which is typically abbreviated to OPO. Once the oscillation threshold is overcome, OPOs are more efficient than sources based on parametric amplification only and usually provide higher output powers. Additionally, only one pump source is needed instead of two for OPA. In principle, also doubly-resonant (signal and idler are feeded back) and pump-enhanced (additional resonator for the pump wave) OPOs exist, but these devices are more difficult to stabilise, tune and adjust [33].

2.2.1 Oscillation threshold

At the onset of parametric oscillation, the assumption of an undepleted pump wave is still valid. Similar to standard laser cavities, an oscillation can start only if the parametric gain overcomes the cavity losses. The signal field at the threshold will therefore be given by

$$E_s(0) = \sqrt{1-V} E_s(L) \ . \tag{2.17}$$

The roundtrip losses V can be caused at the transition from the nonlinear medium to air and vice versa due to imperfect coatings or by residual cavity mirror transmissions. An additional loss mechanism is the linear absorption, which is already included in the gain formula Eq. (2.12). Diffraction losses can be neglected since the diameter of the cavity mirrors is at least one order of magnitude larger than the beam size in our experiments. Rearranging Eq. (2.17), gives the signal gain

$$G = \frac{|E_s(L)|^2}{|E_s(0)|^2} - 1 = \frac{1}{1-V} - 1 \approx V \ , \tag{2.18}$$

showing that the parametric gain equals the losses at the oscillation threshold for $V \ll 1$. This intensity gain depends on $|E_p|^2$ and thus on the pump power P_p. Figure 2.5 illustrates how this threshold is reached once the pump power is high enough for the maximum of the gain to reach the value of the losses. In the idealised case of constant losses, the threshold is first met at vanishing phase mismatch, i.e. $\Delta k = 0$.

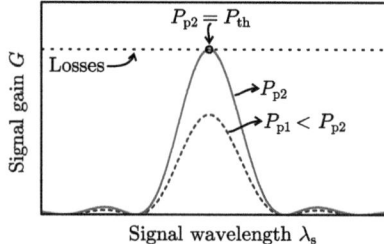

Figure 2.5: Parametric gain G of the signal wave for two different pump powers P_{p1} and P_{p2} with respect to the signal wavelength λ_s. The horizontal dashed line marks the losses. Only the peak of the parametric gain reaches the value of the losses for the higher pump power P_{p2}, therefore P_{p2} meets the threshold condition $P_{p2} = P_{th}$.

Using the condition (2.17) and the solution for the signal gain (Eq. (2.12)), we can determine P_{th}. If absorption is neglected and $V \ll 1$, the equation for the pump threshold P_{th} reduces to

$$P_{th} = \frac{\epsilon_0 c n_p n_s n_i \lambda_s \lambda_i}{8\pi^2 A} \frac{V}{d^2 L^2}. \tag{2.19}$$

Here, A is once more the area of the beam cross-section. This threshold value decreases linearly with lower losses, whereas the dependence on the crystal length L and the nonlinear coefficient d is quadratic. If the crystal length is halved, four times the power is needed to start an oscillation.

In the near infrared with insignificant absorption, the pump threshold P_{th} can easily be calculated. For the common process of converting light of the wavelength 1030 nm to 1500 and 3290 nm radiation in a 5 cm long lithium niobate crystal, the threshold is in the order of 1 W for typical cavity losses of one percent and $d_{IR} = 17$ pm/V.

To generate terahertz waves, however, one needs much higher initial pump powers due to the large idler absorption. Let us assume a process starting at the same pump wavelength in the infrared as before, but now creating an idler wave at a frequency of 1.5 THz, which corresponds to a wavelength of 200 µm. Here, the absorption cannot be neglected and the entire formula of the signal gain Eq. (2.12) has to be considered. Although the nonlinear coefficient should be higher in the THz regime than in the infrared (see Eq. (2.4)), e.g. $d_{THz} = 100$ pm/V, the idler absorption is around 40 cm^{-1} in LiNbO$_3$ [47]. In Sec. 2.3.3, we have already seen, that the resulting gain maximum for 1 W pump power is 0.007 %, which is clearly below cavity losses of 1 %. To overcome this threshold, one would need more than 100 W of pump power, which exceeds the power of standard single-mode and single-frequency pump lasers. This is why continuous-wave parametric terahertz generation has not been demonstrated yet. Therefore, a further enhancement process becomes necessary which will be described in Sec. 2.4 to illustrate our ansatz for terahertz generation.

2.2.2 Oscillation efficiency

In the previous chapters, it is suggested that the highest efficiency is reached for $\Delta k = 0$. The term efficiency is always used as the *quantum* efficiency of a system. Let us now take a closer look at this efficiency η of the parametric process with respect to the pump power. Here, the pump wave can no longer be assumed constant, since it should ideally be depleted entirely. A quantum efficiency $\eta = 100\ \%$ would be reached for full conversion of the pump wave into the signal and idler waves. However, analytic evaluation of the set of all three coupled wave equations (2.3) is not possible. Therefore, in this section we will make the restraining assumptions of vanishing absorptions $\alpha_{p,s,i} = 0$ and perfect phase matching $\Delta k = 0$. Their resulting equations will thus only be valid for the near infrared.

The efficiency η of a parametric process is given by the ratio of the total energy generated at $z = L$ and the incoming pump energy at $z = 0$. Calculations of η for the steady state of an optical parametric oscillator have been performed by Kreuzer and Brunner et al. [49–51]. Based on their analysis, we now estimate the efficiencies for our OPO system. Here, it is assumed that only the signal field is resonant and thus has got the largest amplitude, i.e. $|E_s| \gg |E_p|$. Such a singly-resonant system will be the device used in this work. The conversion efficiency η is given by

$$\eta \equiv 1 - \frac{P_p(L)}{P_p(0)} = \sin^2(\Gamma_B) \ , \qquad (2.20)$$

while the constant Γ_B can be specified through

$$P_p(0)/P_{th} = \frac{\Gamma_B^2}{\sin^2(\Gamma_B)} \ . \qquad (2.21)$$

This result is illustrated in Fig. 2.6. The efficiency η is independent of the pump threshold once it is plotted versus $P_p(0)/P_{th}$. Maximum efficiency is reached for a pump power of $P_p = 2.5 \times P_{th}$. Therefore, it is preferable to operate at this pump power level, to convert as much pump power as possible into the desired signal and idler fields.

The theoretical description bases on the assumption of plane waves with a homogeneous intensity distribution. This might not be valid in most experiments, but this simple model

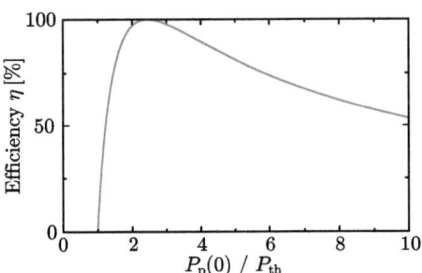

Figure 2.6: *Theoretically predicted efficiency η of optical parametric oscillators with respect to the ratio of incoming pump power $P_p(0)$ and pump threshold P_{th}. The calculations are performed for plane waves with homogeneous intensity distribution [49–51]. The slope of the efficiency is independent of the threshold value P_{th} and η reaches its maximum at $P_p(0)/P_{th} = 2.5$.*

Fundamentals

provides good results in some cases [52]. More elaborate calculations with Gaussian beams can also be found [53], but a qualitative analysis is more descriptive with the simpler model and will be sufficient here. For the Gaussian calculations 100 % conversion efficiency is never reached, since here no perfect overlap can be achieved.

For the experiment, the idler output power is a more important quantity than the efficiency, but those two variables are directly related. The evolution of the theoretical output power of OPOs can be calculated from η via

$$P_i(L) = \frac{\nu_i}{\nu_p} \eta P_p(0) \ . \tag{2.22}$$

The maximum output power of the idler wave, even for 100 % quantum efficiency, is only the fraction ν_i/ν_p of the pump power, which is described by the Manley-Rowe relations [54,55]. Figure 2.7 shows the idler output P_i for different pump thresholds of the standard parametric process from Sec. 2.3.1. The idler power increases monotonously with the initial pump power $P_p(0)$. All these theoretical considerations base on monochromatic waves. Thus, they apply only to continuous-wave OPOs.

In addition, it is necessary to consider the limits of single-frequency operation for such a device to remain in the scope of validity of the theory. Regarding this issue, Kreuzer has made the prediction that single-frequency operation becomes impossible at pump powers above 4.6 times the threshold value [49]. Therefore, all values in Fig. 2.7a are only plotted up to $P_p = 4.6 \times P_{th}$. This implies that for a fixed maximum pump power $P_{p,max}$, there exists an optimum pump threshold P_{th} for achieving the maximum idler output $P_{i,max}$ (see Fig. 2.7b). Consequently, it might not be preferable to reduce the threshold as much as possible, which is usually done in experiments, but to optimise it to maximum output power.

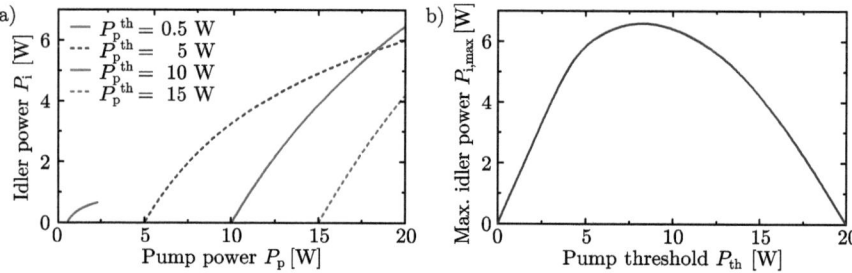

Figure 2.7: a) Theoretically predicted idler power P_i for four different pump thresholds P_{th}. Each curve is plotted up to a pump power of $P_p = 4.6\, P_{th}$ since this is the limit for single-mode operation according to Kreuzer [49]. The maximum pump power taken is 20 W. b) Maximum idler output $P_{i,max}$ for different pump thresholds P_{th} is restricted either by the maximum pump power or the limit of $P_p = 4.6\, P_{th}$.

For the considerations of Kreuzer, only the value of the pump threshold matters, not how P_{th} is achieved. In general, the pump threshold depends on several things, such as the crystal length L and the losses V within the resonator, determining the parametric gain (see Sec. 2.2.1). To adjust the threshold value, one can therefore change either of these parameters: If higher values of P_{th} are needed, shorter nonlinear crystals can be used to lower the gain. This has the advantage that those crystals are easier and cheaper to fabricate. As an alternative, one can employ an outcoupling mirror with a lower reflectivity. This would additionally increase the signal output power which can then also be used. Possible consequences for experiments are investigated in Chapt. 3.

2.3 Quasi phase matching

In the previous section, perfect phase matching was assumed for calculating the idler output. Now, methods to reduce Δk in the experiment are discussed.

If the refractive indices of all three waves were equal, the phase mismatch Δk would vanish, according to the resonance condition from Eq. (2.2). In dispersive media, however, the refractive indices of the interacting waves are not identical. A typical value of Δk for a standard conversion process of light at 1030 nm to approximately 1500 and 3300 nm radiation in lithium niobate is $\Delta k = 2 \times 10^5$ m^{-1}. Therefore, perfect phase matching, i.e. $\Delta k \neq 0$, is not given at all. This results in a coherence length $z = L_c = \pi/\Delta k = 1.5 \times 10^{-5}$ m. After twice the coherence length, the resulting power reaches 0, and thus no amplification takes place [56]. This implies that a disadvantageous choice of the crystal length could result in no output power at all. It is thus very important, to find ways to minimise the phase mismatch for a given crystal dispersion.

In this thesis, we concentrate on the so-called *quasi phase matching*, short QPM [57]. For QPM the nonlinear medium is structured periodically in a way that the spontaneous polarisation $\mathcal{P}$ changes its sign after a certain length. This periodicity Λ corresponds to an additional vector component $K = 2\pi/\Lambda$ fulfilling

$$\vec{k}_\text{p} = \vec{k}_\text{s} + \vec{k}_\text{i} + \vec{K} \quad \Rightarrow \quad \Delta k = 0 \tag{2.23}$$

as illustrated in Fig. 2.8. If the period length Λ of this structure is adjusted correctly, a nonlinear process can take place efficiently over the entire crystal length. Thus, the QPM period Λ is related to the coherence length via $\Lambda/2 = L_c$ for phase matching. For the wavelength example from Chapt. 2.1.2, a periodicity $\Lambda = 30$ µm is needed to compensate the mismatch. Such a structuring makes efficient conversion possible over several centimeters instead of micrometers of crystal length.

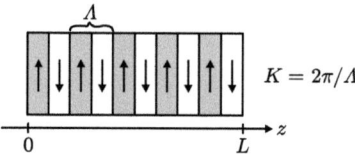

Figure 2.8: *Periodically oriented crystal with a structure period size of Λ and the crystal length L, providing an additional vector component with the absolute value $K = 2\pi/\Lambda$. The arrows indicate different orientations of the spontaneous polarisation.*

The only drawback of this method is a resulting reduction of the absolute value of the nonlinear coefficient: $d \rightarrow d_\text{eff} = 2d/\pi$ [56]. This loss is due to the rectangular modulation instead of a perfect sinusoidal one which cannot be achieved experimentally in nonlinear crystals. The factor $2/\pi$ originates from the Fourier analysis [56]. For ferroelectric crystals like lithium niobate, such a structuring can be achieved by periodically inverting the direction of the spontaneous polarisation via applying strong electric fields. This process is called *periodic poling* [58]. The common abbreviation for periodically poled lithium niobate is PPLN. One period length Λ is then defined as the length of two oppositely inverted regions next to one another (see Fig. 2.8).

In general, almost arbitrary orientations of the wave vectors as well as to the grating vector are possible. If the vector sum of all contributions is small, i.e. $|\Delta k| \leq 2\pi/L$, then the phase matching condition is fulfilled (see Eq. (2.23)). However, it is favourable, that the wave vectors are collinear as done in this work, because this gives a larger interaction length leading to higher conversion efficiencies. In this thesis, all waves propagate on the z-axis and thus only the sign of the vectors is of importance.

One should keep in mind that the fundamental resonance condition Eq. (2.2) in combination with the phase matching Eq. (2.23) is not unique even for a fixed poling period. There are two equations for three ν_l and three different refractive indices n_l:

$$\nu_p = \nu_s + \nu_s \quad \text{and} \quad \frac{n_p \nu_p}{c}\vec{e}_p = \frac{n_s \nu_s}{c}\vec{e}_s + \frac{n_i \nu_i}{c}\vec{e}_i + \frac{1}{\Lambda}\vec{e}_K \ .$$

Here, in the phase matching condition the wave vectors are rewritten by their frequencies $k_l = 2\pi n_l \nu_l / c$ and unity vectors $\vec{e}_l$, which in our case point only into $\pm z$ directions. Although frequencies and refractive indices are not independent of one another, two equations are not sufficient for determining all values uniquely. It is thus possible that several parametric processes can be phase matched by the same QPM period. This will be illustrated further in the following sections, in which parametric processes for terahertz generation are presented.

2.3.1 Standard forward parametric process

The simplest phase matching scheme for periodically oriented materials is the one with all vector components pointing into the same direction as illustrated in Fig. 2.9. Here, the sum of the signal and idler wave vectors together with the grating vector gives the pump wave vector. Our type of phase matching with all waves being extra-ordinarily polarised is called type 0 phase matching.

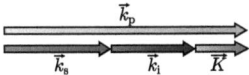

Figure 2.9: *Quasi phase matching provides an additional vector component $\vec{K}$ to the signal $\vec{k}_s$ and idler wave vectors $\vec{k}_i$, making their sum equal to the pump wave vector $k_p = k_s + k_i + K$.*

Varying the period length allows a tuning of the frequencies of the generated waves. Figure 2.10 shows this tuning for the case of $\lambda_p = 1030$ nm in MgO-doped LiNbO$_3$ at room temperature. The two branches of the tuning curve correspond to the signal and the idler wavelengths, respectively. The turning point at $\Lambda = 31.48$ µm, where the signal and idler frequency are the same, is called point of degeneracy.

Additional tuning can be achieved by changing the temperature of the nonlinear medium. Since the refractive index n is temperature dependent, this alters the phase matching

FUNDAMENTALS

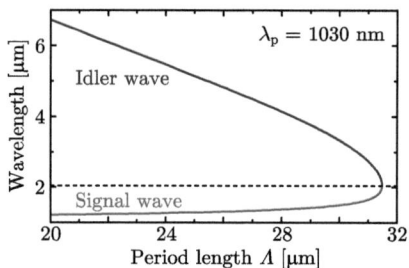

Figure 2.10: *Wavelength tuning of an optical parametric oscillator with respect to the grating period Λ at room temperature. The pump wavelength λ_p is fixed to be 1030 nm. Above the turning point at $\lambda = 2060$ nm ($\Lambda = 31.48$ µm), the wavelengths belong to the idler λ_i (blue line), below this point to the signal wavelength range λ_s (red line).*

condition and thus the resulting wavelengths. A change in temperature also slightly affects the poling period, through thermal expansion, which has to be taken into account for predicting the exact tuning behaviour [59].

If the pump wave with a wavelength $\lambda_p = 1030$ nm is sent into a PPLN crystal with a period length of $\Lambda = 30$ µm at room temperature, the resulting wavelengths are 1520 nm (signal) and 3190 nm (idler). All these wavelengths are in the transparency range of lithium niobate. Therefore absorption in the parametric gain can be neglected and their refractive indices are given by the Sellmeier equation of Gayer et al. [46]. With the technique of quasi phase matching the effective nonlinear coefficient for this process becomes approximately 17 pm/V [60] instead of almost 30 pm/V, which is why the smaller value was used in earlier example calculations.

2.3.2 Forward parametric process into the terahertz regime

Why do we need a different phase matching scheme for terahertz generation in lithium niobate? By looking at the wave vector equation (2.23), one can see that the standard phase matching scheme is sufficient as long as the refractive indices of the interacting waves are similar. All vectors, $\vec{k}_l$ and $\vec{K}$, can only be parallel if n_p is the largest refractive index. In the terahertz range, however, the refractive index is in the order of five or even higher instead of approximately two for the infrared waves [46, 47]. Therefore, phase matching with all vectors, $\vec{k}_l$ and $\vec{K}$, pointing into the same direction no longer works. As a consequence, one can look at processes where one vector component is anti-parallel to the rest.

The direction of the grating vector $\vec{K}$ is not defined a priori. Even if all wave vectors $\vec{k}_l$ are chosen to be parallel, $\vec{K}$ can flip sign (see Fig. 2.9). Such a process can be used

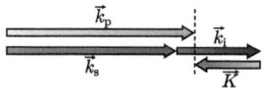

Figure 2.11: *Forward parametric terahertz process. Phase matching is achieved by $\vec{K}$ being antiparallel to pump $\vec{k}_p$, signal $\vec{k}_s$ and idler $\vec{k}_i$ wave vectors: $k_p = k_s + k_i - K$.*

20

for terahertz frequency generation. If we take the same period length as in the example above, the resulting terahertz frequency for a pump wave at a wavelength of 1030 nm would be 3.1 THz. It should be emphasised that this process could occur in addition to the standard one from above (Sec. 2.3.1) within the same crystal structuring if the high threshold due to absorption is overcome.

2.3.3 Backwards parametric process into the terahertz regime

Lets us now assume that one of the wave vectors, e.g. $\vec{k}_i$, is anti-parallel to the other participating waves, as displayed in Fig. 2.12. Such a backwards process was proposed in 1966 by Harris [61], and this process has already been used experimentally for terahertz generation by Yu et al. in 2007 in pulsed systems [62].

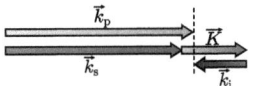

Figure 2.12: *The idler wave $\vec{k}_i$ travels backwards with respect to pump $\vec{k}_p$ and signal wave vectors $\vec{k}_s$. The grating vector contribution $\vec{K}$ is parallel to $\vec{k}_p$ and $\vec{k}_s$: $k_p = k_s - k_i + K$*

Phase matching is also achieved with a poling period of $\Lambda = 30$ µm, resulting in 1.3 THz radiation being generated by light of 1030 nm pump wavelength. Although the same pump wavelength and QPM period are assumed for both, the forward and backward parametric process for terahertz waves, the resulting THz frequencies differ by more than a factor of two.

So far, we have only considered collinear waves to discuss different parametric processes but in general also other situations are possible. The three waves and the grating structure could be at almost arbitrary angles. One simple case would be, that pump and signal wave remain parallel, the crystal grating vector can be slanted such that the resulting idler wave is emitted perpendicular to $\vec{k}_p$ [63].

2.4 Terahertz wave generation with cascaded parametric processes

To extend the tuning range of a continuous-wave OPO to the terahertz range, we begin with the standard OPO process, converting a pump wave with a wavelength of 1030 nm to a signal wave at a wavelength around 1500 nm (see Sec. 2.3.1). Due to the large absorption of terahertz waves in lithium niobate [47], high pump powers are needed for starting parametric processes, which can be overcome by means of resonant enhancement. At first glance, it looks straight forward to build an extra resonator to increase the pump power itself. However, this makes the system more complicated and difficult to stabilise [33]. These issues can be avoided since we already have one resonant wave, the signal wave, whose power can grow up to the kilowatt level for initial pump powers P_p in the order of 10 W [64]. Therefore, the signal wave can act as a pump wave for a secondary parametric process, $\nu_\mathrm{s1} = \nu_\mathrm{s2} + \nu_\mathrm{i2}$, as displayed in Fig. 2.13. We call this secondary process a *cascaded parametric process* and this one actually is capable of overcoming the threshold for terahertz generation.

The wave vector equation (see Sec. 2.3) for this cascaded process is given by

$$k_\mathrm{s1} = k_\mathrm{s2} - k_\mathrm{i2} + K \ . \tag{2.24}$$

This illustrates that the terahertz wave is traveling backwards with respect to the other interacting waves in order to fulfill the phase matching condition.

In Sec. 2.3.3, we have already seen that an infrared and a terahertz process can be phase matched within the same QPM period. This still holds if the initial wavelength of the terahertz process is 1500 nm instead of 1030 nm, since Λ is mostly determined by the strong difference between n_IR and n_THz. With a frequency separation of just some terahertz both signal waves, ν_s1 and ν_s2, can be easily captured within the same cavity. This device thus bases on a pump-enhanced process with the second pump wave at ν_s1 being

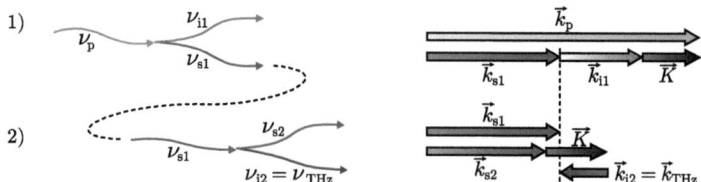

Figure 2.13: *Scheme of cascaded parametric processes. 1) A pump wave with a frequency of ν_p generates a signal at ν_s1 and an idler wave at ν_i1 in a standard forward process. The right side shows the corresponding phase matching condition $k_\mathrm{p} = k_\mathrm{s1} + k_\mathrm{i1} + K$. 2) The signal wave ν_s1 from the first process acts as a pump wave for an additional parametric process, generating a second signal ν_s2 and idler wave ν_i2. This second idler wave is in the terahertz frequency range $\nu_\mathrm{i2} = \nu_\mathrm{THz}$. Here, a backward phase matching scheme is used: $k_\mathrm{s1} = k_\mathrm{s2} - k_\mathrm{i2} + K$.*

generated within the resonator and therefore automatically selecting a cavity mode. No careful impedance matching is necessary, avoiding the complicated stabilisation issues of multiply-resonant OPOs.

Chapter 3

Output power optimisation

Due to the large absorption of terahertz waves in nonlinear crystals, high powers are essential for terahertz generation. This chapter therefore deals with power optimisation of singly-resonant optical parametric oscillators.

First of all, the experimental methods necessary for studying a standard OPO are introduced. In Sec. 2.2.2, it has been shown that the maximum single-mode idler output power depends on the pump threshold P_{th}. This value of P_{th} varies with losses as well as with the parametric gain while the latter one is a function of the crystal length L. In the following sections, it is now analysed how the idler power can be optimised by either changing the L or the cavity losses V. High idler powers also correspond to high signal powers, which are needed for the cascaded terahertz process (see Sec. 2.4). The efficiency of the parametric process is investigated experimentally and compared with the theory predictions from the previous chapter.

3.1 Experimental methods for the infrared setup

The first continuous-wave (cw), singly-resonant optical parametric oscillator which relies on lithium niobate crystals was demonstrated in 1996 by Bosenberg et al. [65]. The following sections introduce the required experimental components for such a system. First, the nonlinear lithium niobate crystals and their specifications are presented. Afterwards, the setup of a standard OPO, emitting infrared waves, is explained, including ways of measuring the spectral features and experimentally determining the efficiency of such a device.

3.1.1 Nonlinear crystals

In this work, magnesium-oxide-doped lithium niobate crystals provided by HC Photonics Corp. are used. Their MgO content is specified to be more than 5 wt% in the melt for the crystal to withstand optical damage [66]. All crystals have a height of 0.5 mm. They are positioned on an oven made of copper, which can provide temperatures up to 180 °C. The temperature stability is ±0.01 K, measured with a PT1000 being in thermal contact with the copper block. This oven temperature is assumed to be the crystal temperature T in all following chapters, since the crystal is covered with a copper lid and after changing the temperature the setup is given enough time to relax into thermal equilibrium.

Crystal set I: Five different crystal lengths are employed: 1.5, 1.7, 2.0, 2.5 and 5.0 cm. The crystals have got seven differently poled sections with period lengths $\Lambda = 28.5$ to 31.5 µm and 0.5 µm period increments for every crystal length L. This crystal design is displayed in Fig. 3.1. The light and dark grey shaded regions in the crystal symbolise the two distinct orientations of the crystallographic symmetry axis (see Sec. 2.3).

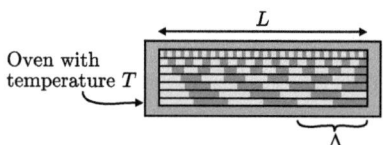

Figure 3.1: *Nonlinear crystal of length L on an oven with the temperature T. The crystal is divided into seven sections with different QPM periods Λ.*

The end surfaces of all samples are anti-reflection coated for pump, signal and idler wavelengths. In particular, the coatings for the resonant signal waves have to be sufficiently good in order to minimise cavity losses. Crystals with $L = 5$ cm or $L = 2.5$ cm have got coatings specified with a residual reflectivity $R \leq 1$ % at (1500 - 1600) nm and a minimum of 0.3 % at 1600 nm, while the other crystal lengths are characterised with $R \leq 1$ % at (1500 - 1600) nm and a minimum of $R = 0.1$ % at 1550 nm.

Crystal set II: To extend the tuning range of the OPO, additional crystals of 5 cm length (also from HC Photonics Corp.) are available. Here, the QPM periods are 24.4, 24.8, 25.0, 25.3, 25.6, 26.0 and 26.4 µm on each crystal. These crystals are anti-reflection coated for different wavelengths accordingly. The minimum reflectivity is achieved at 1300 nm.

3.1.2 Optical parametric oscillator for near and mid infrared waves

Experimental Setup

The pump source for the OPO is a continuous-wave Yb:YAG disc laser from the ELS GmbH, emitting at a wavelength of $\lambda_\text{p} = 1030$ nm with a specified maximum power of $P_\text{p,max} = 20$ W. This power can be continuously regulated in the experiment via a combination of a half-wave-plate with a polarising beam splitter. The linewidth of the laser light is below 5 MHz and the beam profile is Gaussian ($M^2 \leq 1.1$). A Faraday isolator in front of the laser prevents light from being reflected back into the laser cavity. Afterwards, the pump light is focussed into the lithium niobate crystal such that the radius of the focal spot at the crystal centre is about 100 µm, defined by the points where the intensity is reduced to $1/e^2$ of its maximum value. The corresponding experimental setup is displayed in Fig. 3.2.

The central component of a parametric oscillator is its cavity containing the nonlinear medium. In this thesis, a bow-tie configuration is used, consisting of two plane mirrors (OP1 and OP2, standing for OPO plane mirror) and two curved ones (IC and OC, input and output coupler, curvature radius 100 mm). All these dielectric mirrors are highly reflecting $R_M \geq 99.9$ % in the signal wavelength range between 1400 and 1800 nm and anti-reflection coated for the pump and idler wavelengths. One of the plane mirrors (OP2) can be replaced by outcoupling mirrors with residual transmission for the signal wave of 0.5 or 1.5 % around 1520 nm to adjust the cavity losses and to vary the output signal power. For the crystals with poling periods 24.4 to 26.4 µm, a different set of highly reflecting mirrors is used with the maximum reflectivity around 1300 nm in analogy to the crystal coatings.

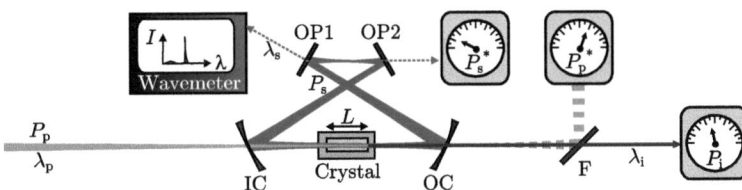

Figure 3.2: *Schematic setup of the optical parametric oscillator. Its cavity comprises two curved mirrors (IC, OC) and two plane ones (OP1, OP2). Here, λ_p, λ_s and λ_i denote the wavelengths of the three interacting waves pump, signal and idler. Their powers are labeled by P_k ($k \in \{p, s, i\}$), with P_k^* signifying the residual power outside the resonator.*

The idler wave experiences losses before reaching its detector due to a residual reflectivity at the end facet of the crystal (only $R < 10$ % are specified) and the imperfect transmission of OC ($R \approx 95$ %). In addition, the idler wave needs to be separated from the pump wave with a dielectric calcium-fluoride mirror F. Although F is highly reflecting (at least 99.9 %) for the pump wave, the transmitted idler wave is reduce by approximately 10 to 15 % depending on the wavelength. Since these specifications are not precise enough, we do not correct measured idler powers to these attenuations.

The optical path length of the resonator is optimised with respect to the different crystal lengths L. This can be calculated by using the ABCD matrix formalism to estimate the beam radius on its path through a ring resonator [67, 68]. We use approximately 43 to 41 cm of cavity length, adjusted to each L individually. The $1/e^2$-radii of the resulting focal spots of the signal wave at the crystal centre range from 80 to 90 µm to ensure an overlap of pump and signal waves.

Measurement techniques

To characterise the performance of the infrared OPO, the output powers need to be measured.

One way of determining the conversion efficiency η of the parametric process, is to measure the pump depletion. For this purpose, the residual pump power behind the cavity with ongoing oscillation $P_p^*(L)$ is detected and compared with the pump power for a blocked resonator $P_p^*(0)$ (see Fig. 3.3). Although $P_p^*(0)$ is slightly smaller than the initial pump power at the front facet of the crystal because of losses caused by the crystal and mirror surfaces, this technique features that both powers are measured at the same position and thus pass through exactly the same optical components. Therefore, their ratio

$$\eta = 1 - \frac{P_p^*(L)}{P_p^*(0)}, \qquad (3.1)$$

can be well used to calculate the efficiency η. Nevertheless, all losses of the pump wave are small, since all components are anti-reflection coated for the pump wave. We therefore assume in the following that the measured value of $P_p^*(0)$ is also the pump power entering the crystal, i.e. $P_p^*(0) = P_p(0)$. All powers are measured with thermal detectors from the Coherent GmbH, whose rise time is two seconds, and their absolute accuracy is specified to be $\pm 1\ \%$.

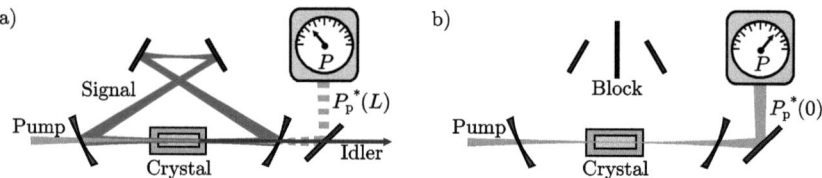

Figure 3.3: *Scheme for determining the pump depletion. The pump power behind the resonator is measured with a power meter (labeled with P), once with oscillating signal wave $P_p^*(L)$ (a) and then with a blocked cavity $P_p^*(0)$, preventing an oscillation (b).*

An additional way of analysing the efficiency of the OPO is by its idler output power P_i. To measure P_i, it has to be separated from the residual pump power. The signal power within the resonator can be deduced from the power measured outside the cavity P_s^* by assuming a known mirror transmission. However, this only works for the outcoupling mirrors, since here the residual transmission can be determined precisely enough. Therefore, the signal power within the cavity is given in arbitrary units for the high-finesse cavity.

For characterising the potential performance of an optical parametric oscillator, the highest possible output powers for a certain pump power should be employed for comparison with theory because experimental imperfections can only reduce but not enhance the idler power. Therefore, for each value of P_p several idler powers are recorded and the highest value is taken as the data point. It is also important to wait a couple of minutes once the pump power has been changed for the OPO to reach its equilibrium state.

In addition, spectral features, such as wavelength values, linewidths and tuning characteristics, are of great interest. To analyse the spectral properties of the signal waves, a Burleigh wavemeter WA-1500 in combination with a Burleigh spectrum analyser WA-650 is used (see Fig. 3.2). A precision of ± 0.2 ppm, corresponding to ± 40 MHz at 1600 nm, and a resolution of 4 GHz are achieved. Analysis software enables continuous tracking of the maxima of the resonant waves. The residual signal power of a few milliwatt, leaking through the highly reflecting OPO mirrors, is sufficient for this analysis. For faster but less accurate wavelength measurements, an Agilent spectrum analyser can be employed. Its accuracy is specified to be ± 0.5 nm, and the resolution is 0.06 nm which corresponds to 7 GHz at 1600 nm.

To confirm single-frequency operation of pump or signal waves, we use Fabry-Pérot-interferometers (FPI) with a free spectral range of 1.5 GHz and a finesse of 200. One the one hand, such a FPI signal can be used to observe the stability of the signal frequencies while on the other hand, it can be employed within an active stabilisation system. Even if the OPO runs without modehops, the wavelength of the resonant signal wave can change slightly due to a drift of the cavity. To compensate for this change, the FPI data can be used as a frequency reference, locked to the highest detected peak. Via a computer programme any deviation from this fixed position is tracked and translated into a voltage at a piezo element to move one of the plane cavity mirror, in order to compensate this deviation. This piezo element allows a fine adjustment of the cavity length and can move the mirror up to 5 µm in total. Such an active stabilisation is very useful for longer measurements.

OUTPUT POWER OPTIMISATION

3.2 Efficiency characterisation of the parametric oscillator

OPOs are working horses for spectroscopy. For applications such as high-resolution spectroscopy, a single-frequency system is essential. Therefore, in this chapter, idler powers are recorded only as long as the OPO operates in single-mode status (see Sec. 3.1.2). Figure 3.4 shows the running parametric oscillator exemplary for a 5-cm-long lithium niobate crystal, covered with a lid. This crystal can be exchanged with others of different lengths (2.5, 2.0, 1.7 and 1.5 cm, see Sec. 3.1.1) as analysed in Sec. 3.2.1. Further on as done in Sec. 3.2.2, the plane mirror OP1 (R > 99.9 %) can be replaced by an outcoupling mirror, R = 99.5 or R = 98.5 %, to vary cavity losses.

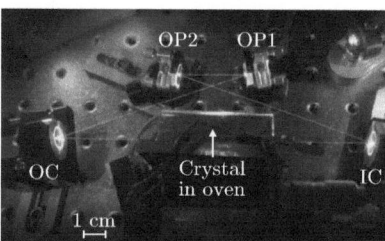

Figure 3.4: *Photograph of an operating OPO with a 5-cm-long nonlinear crystal, situated in an oven. The path of the resonating wave is visible due to the non phase matched frequency doubled light of the signal wave (orange) which illustrates the bow-tie configuration. The cavity mirrors are labeled according to their introduction in Chapt. 3.1.*

For the following measurements a poling period of $\Lambda = 30.0$ µm is used, with signal wavelengths around 1520 nm resulting from a pump wavelength of 1030 nm, leaving absorption of all interacting waves negligible. Idler and signal powers outside the cavity are measured in combination with the pump depletion to determine the efficiency of the OPO (see Sec. 3.1.2).

3.2.1 Different crystal lengths

In this section, the output idler powers with respect to the pump power entering the crystal are measured for all different crystal lengths. To keep the cavity losses fixed, the smaller crystals with $L = 2.0$ cm or below are operated with one outcoupling mirror with a residual transmission of 0.5 %, since the coatings of these crystals are better. This leaves a total amount of resonator losses – a combination of mirror transmission and residual crystal reflectivities – of 0.7 % for all crystals. This causes threshold values for the parametric process ranging form 0.95 to 10.5 W.

The resulting output powers are displayed in Fig. 3.5. One limitation of the output is the maximum pump power available of $P_{\text{p,max}} = 18.8$ W, actually reaching the crystal front facet. For crystal lengths of $L = 1.5$, 1.7 and 2.0 cm no multi-mode operation occurs until this maximum pump power is reached. Consequently, the ratio of P_p/P_th stays

Output power optimisation

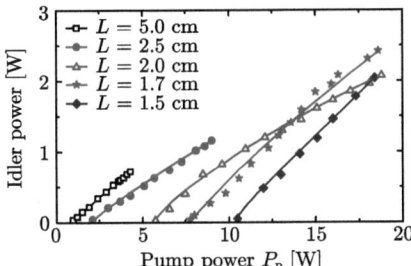

Figure 3.5: *Idler output power with respect to the initial pump power for five different crystal lengths L. Measurements are taken only as long as the OPO output is single-frequency or the maximum pump power is reached. The highest maximum idler output is achieved for L = 1.7 cm. Solid lines act as guides to the eye.*

below Kreuzer's limit for all of these crystals. The OPO with a 2.5-cm-long nonlinear element oscillates on only one mode until $P_p/P_{th} = 4.3$. For a 5-cm-long crystal this single-frequency limitation is reached at 4.5 times the pump threshold.

The errors due to the uncertainty of the power meter of ± 0.01 W are smaller than the symbol size, and thus not visible in the graph. The reproducibility of each data point is estimated to be ± 10 %. For every pump power, several idler and the corresponding residual pump powers are measured and the one with the highest idler value is taken since this shows, how well the OPO is able to perform (see Sec. 3.1.2).

A shorter crystal length leads to a higher pump threshold (see Eq. (2.17)). For the different lengths, the corresponding threshold values P_{th} and the maximum idler output powers $P_{i,max}$ are collected in Tab. 5.2. The highest single-frequency idler output of $P_{i,max} = 2.42$ W is achieved for a $L = 1.7$ cm with a pump threshold of 7.8 W. This threshold value corresponds to a fraction of the maximum initial pump power of $P_{p,max}/P_{th} = 2.41$.

L [cm]	P_{th} [W]	$P_{i,max}$ [W]
5.0	0.95	0.72
2.5	2.12	1.16
2.0	5.74	2.07
1.7	7.81	2.42
1.5	10.5	2.04

Table 3.1: *Measured pump threshold values P_{th} and maximum idler output powers $P_{i,max}$ for five different crystal lengths L.*

The measured idler powers can be used to calculate the efficiency of the OPO. Here, 100 % quantum efficiency are achieved for an idler output power of $P_i = P_p \nu_i / \nu_p$, as described by the Manley-Rowe relations (see Eq. (2.22)). Another way of determining the efficiency η of the parametric process is via the pump depletion (see Sec. 2.2.2). This depletion is also measured using the five different crystal lengths and varying the pump power. Figure 3.6

OUTPUT POWER OPTIMISATION

compares the characteristics of the efficiency, as evaluated from the idler powers, and the pump depletion for a 5-cm-long crystal, exemplarily. The efficiency given by pump depletion is higher than the other one from P_i. Between the threshold value and $2.5 \times P_{th}$ it rises to 60 %. Above this value a smaller increase can be seen. The maximum efficiency achieved is 70 % at four times above the pump threshold, after which η decreases slightly. From the idler power, the highest deduced efficiency value is $\eta = 55$ % at $P_p = 4.55 \times P_{th}$, while the behaviour is similar to the one observed via the pump depletion. The idler efficiency is lower than the pump efficiency since the idler power reaching the detector is lowered while passing through components by approximately one fourth (Sec. 3.1.2).

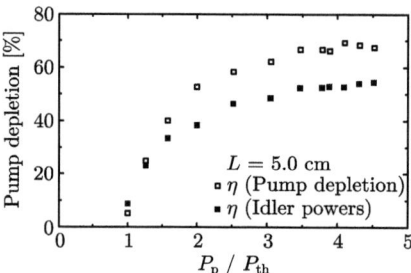

Figure 3.6: *Measured efficiencies of the parametric process deduced from pump depletion and idler powers exemplary for a 5-cm-long crystal. The abscissa shows the initial pump power at the front facet of the crystal P_p, normalised to the pump threshold value P_{th}. The maximum efficiency reached is 70 %.*

3.2.2 Varying cavity losses

In contrast to section 3.2.1, we now keep the crystal length fixed at 5 cm and vary the cavity losses by using different outcoupling mirrors. The resulting idler powers are displayed in Fig. 3.7. Once more, only data points with single-frequency operation are recorded. If all mirrors are highly reflecting, the pump threshold is 0.95 W as shown in the section above with the same single-frequency limit of $P_p/P_{th} = 4.5$. One plane mirror is then replaced by an outcoupling mirror. For a residual transmission of 0.5 %, the resulting threshold becomes 2.4 W and a maximum idler power of 1.37 W at $4.3 \times P_{th}$

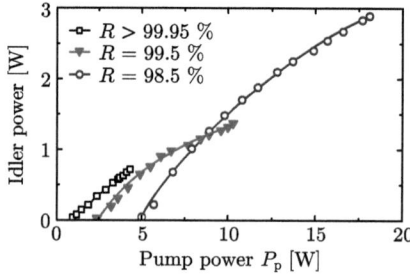

Figure 3.7: *Measured idler powers for three different cavity losses and 5 cm crystal length. For the lowest pump threshold all mirrors are highly reflecting. The higher thresholds are generated by replacing one plane mirror with an outcoupler with residual transmission of 0.5 or 1.5 %, respectively. Solid lines act as guides to the eye.*

is observed. If the residual transmission is 1.5 %, the pump threshold increases to 5.0 W and the idler power grows up to $P_{i,max} = 2.89$ W.

For the highest idler output, the efficiency is plotted in Fig. 3.8. Here, the maximum pump power corresponds to 3.6 times the pump threshold. The highest efficiency of 73 % is reached at $P_p/P_{th} = 2.56$. After this value, η remains almost constant around 71 %. The measured maximum idler efficiency of 65 % is once more below the efficiency determined by the pump depletion while the general characteristics of both efficiencies reveal a very similar behaviour, as observed also in Fig. 3.6.

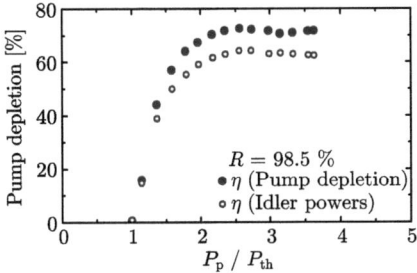

Figure 3.8: Measured pump depletion and idler efficiency for a 5 cm long crystal and one outcoupling mirror with a reflectivity of $R = 98.5$ %. All other mirrors are highly reflecting. The x-axis displays the initial pump power P_p, normalised to the pump threshold value $P_{th} = 5$ W. The highest pump depletion of 73 % is reached at $P_p/P_{th} = 2.56$.

Simultaneously to the idler output, the signal power outside the resonator can be measured. This power increases with higher residual transmission of the outcoupling mirror. Signal power characteristics for the outcoupling mirror with $R = 98.5$ % are displayed in Fig. 3.9. The maximum signal power is 7.63 W in addition to 2.89 W of outcoupled idler power (see Sec. 3.2.1). The total amount of outcoupled power is thus approximately 10.5 W, corresponding to almost 60 % of the incoming pump power which is as good as the measured idler efficiency (see Fig. 3.8).

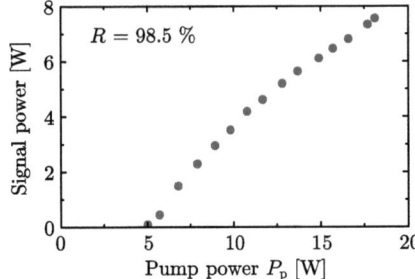

Figure 3.9: Signal power outside the cavity for a 5-cm-long crystal measured behind an outcoupling mirror with a reflectivity of $R = 98.5$ %,. The pump threshold is 5 W as already shown in Fig. 3.7. The highest signal power achieved is 7.63 W for the maximum pump power $P_{p,max}$ used.

OUTPUT POWER OPTIMISATION

3.3 Discussion of performance optimisation

In section 2.2.2, the theoretical expectations for the efficiencies and output powers of standard optical parametric oscillators were introduced. Now, we can compare these predictions with the measured values presented above.

3.3.1 Single-mode operation

L. B. Kreuzer states, that single-frequency performance is no longer possible above $P_\mathrm{p} = 4.6 \times P_\mathrm{th}$ [49]. The single-frequency operation limits measured with our setup are between 4.3 and 4.5 times of the pump threshold. Thus we have shown experimentally that Kreuzer's limit applies to our set of wavelengths.

Other groups have seen different limits for multimode operation in their setups. For example, Zaske et al. [69] report a single-frequency limit of twice the pump threshold being more than a factor of two below Kreuzer's prediction of $P_\mathrm{p} = 4.6 \times P_\mathrm{th}$. In that experiment a pump laser at a wavelength of 532 nm is used, generating signal waves between 1400 and 1450 nm. Such a system would of course need a different idler output power optimisation. Kreuzer's limit is not as universal as claimed, since the group velocity dispersion of the pump wave is not include [49]. Therefore, the limit for single-mode operation should depend on the wavelengths involved in the parametric processes. It is thus a coincidence that Kreuzer's prediction fits to our experimental parameters.

As a future perspective, one could think of using only a thin plate of lithium niobate in combination with a kW fibre pump laser. Yet, one has to bare in mind, that the pump threshold depends quadratically on the length for vanishing absorption and thus L cannot be reduced by the same factor with which $P_\mathrm{p,max}$ is increased.

3.3.2 Efficiency of the parametric process

The plane-wave model can be used for comparing the efficiency of the parametric process with our measured values (see Sec. 2.2.2). This theoretical efficiency shows the same behaviour for all pump thresholds if plotted against the normalised pump power $P_\mathrm{p}/P_\mathrm{th}$ (see Fig. 3.10, solid line). In addition, a more elaborate theory of Bjorkholm et al. [53] based on Gaussian beams is considered (Fig. 3.10, dashed line). These efficiencies are compared with the measured values as determined by the pump depletion. One can clearly see, that the optimum of 100 % quantum efficiency, as predicted by the plane-wave model, is not reached.

In general, the measured values lie below η from the simpler model while they are mostly higher than the ones given by Bjorkholm's theory. The behaviour of the high-finesse cavity efficiency (open squares) is more similar to the Gaussian beam model, while the data of the lower finesse cavity is closer to the curve derived from the plane-wave model although no deliberate changes in focussing or shaping of the pump and signal waves is

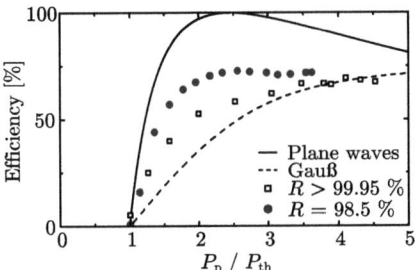

Figure 3.10: *Comparison of the theoretically expected quantum efficiencies (solid line for plane waves and dashed line for Gaussian beams) with the measured pump depletion. Both measurements are performed with a 5-cm-long crystal. Once all mirrors are highly reflecting (black squares) and once one outcoupling mirror with a reflectivity of 98.5 % is used (blue dots).*

performed. The curve of the high finesse cavity rises slower and reaches its maximum at $P_p = 4 \times P_{th}$, while the other measurements reach their maximum at $P_p = 2.4 \times P_{th}$.

It has been shown experimentally by Bosenberg et al. that efficiencies as high as 93 % can be achieved in singly-resonant cw OPOs based on lithium niobate [52], agreeing nicely with the plane wave theory. This higher value with respect to our measurements could be explained by different adjustment of the pump focus. Yet, Bosenberg's results indicate that the simpler model, although assuming plane waves, can be used for Gaussian beams in parametric oscillators. The maximum efficiency, predicted by the more elaborate theory of Bjorkholm et al. [53], is only 70 %. However, theoretically expected efficiencies should rather be higher than the measurement and not lower. This justifies comparing the idler output powers with those predicted by the simpler model.

3.3.3 Idler power optimisation

Figure 3.11 shows the measured idler powers in comparison with the theoretically calculated values with respect to the normalised pump power P_p/P_{th}. For each measurement series the pump power is divided by the individual threshold value of this series, $P_{th,1}$ to $P_{th,5}$ being 0.9, 2.1, 5.7, 7.8 and 10.5 W, respectively. The theory curves are adjusted to match the measured pump thresholds, but no other fit parameter is inserted (see Sec. 2.1.3).

The general behaviour of the measured idler powers corresponds well to theoretical expectations. In particular, the highest achievable idler output for this OPO configuration is predicted and experimentally reached for a threshold value of 7.8 W, corresponding to a crystal length of $L = 1.7$ cm. This threshold is related to the maximum pump power by $P_{p,max}/P_{th,4} = 2.4$. The limit of Kreuzer [49] for single-frequency operation entails that the maximum quantum efficiency is obtained for $P_p/P_{th} = 2.5$, which agrees very well with this experimental data. Further evaluation can be found in Fig. 3.12. Here, the theoretically expected maximum idler powers are plotted versus the pump threshold and compared with the experimental results. Once more, the slope of measurement and theory match, but the absolute values differ.

OUTPUT POWER OPTIMISATION

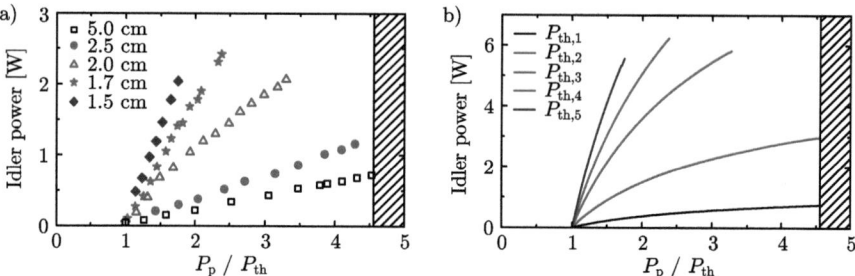

Figure 3.11: a) Measured idler output powers for five different crystal lengths L. b) Theoretically calculated idler powers for five different pump thresholds $P_{th,1}$ to $P_{th,5}$ which correspond to the measured threshold values. Both abscissae show the pump power, normalised to the individual pump threshold values of each curve.

The model predicts a maximum output power of more than 6 W. In the experiment, approximately half this value is measured (see Fig. 3.11). One way of explaining this could be the point that the theory bases on plane waves but Gaussian waves are used in the experiment and that a conversion efficiency of $\eta = 1$ is not reached. According to our results, the highest output powers are achieved for either a crystal length of 1.7 cm and only highly reflecting mirrors or a 5-cm-long crystal with one outcoupling mirror (R = 98.5 %) for a maximum pump power of 18.8 W. Shorter crystals have the advantage that they are cheaper to fabricate and easier to periodically structure with long range order over the entire crystal length. Outcoupling mirrors in contrast provide high signal powers in addition to the idler powers outside the cavity. However, for certain applications a lower threshold will still be favourable because smaller pump lasers can be used. Here, long crystals and highly reflecting mirrors can be employed. Alternatively, one can place a gain medium into the cavity to lower the threshold [70].

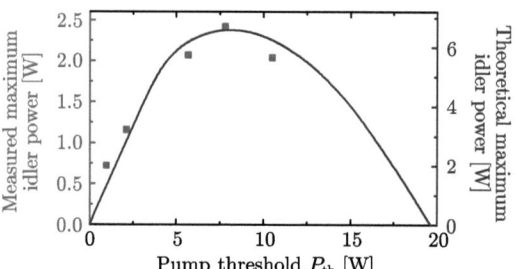

Figure 3.12: Measured maximum idler output power $P_{i,max}$ (red squares) with respect to the pump power threshold P_{th}. The blue solid line shows the theoretically expected values of $P_{i,max}$ for a maximum pump power of 19 W.

36

To conclude one can say, that for each application the setup design has to be determined individually according to the needs of the experiment and sometimes shorter crystals can be better suited.

CHAPTER 4

SPECTRAL FEATURES OF THE RESONANT WAVES

Optical parametric oscillators are widely used in spectroscopy because of their large tuning ranges. Generally, tuning can be achieved by varying the poling structure and the crystal temperature. In this chapter, emphasis is placed on the spectra of the resonant waves instead of the output powers.

The first section deals with the standard parametric process, pumped by a wavelength of 1030 nm. At high pump powers, additional features occur in the spectra and characteristic difference frequencies can be observed which are interpreted here in the second section. Some of these processes can be used for terahertz generation as discussed in the last section.

The experimental methods introduced in the previous chapter also apply to these measurements, the same nonlinear crystals and standard OPO setups are used. To determine the tuning range, all QPM periods available are employed , one type comprising 28.5, 29.0, 29.5, 30.0, 30.5, 31.0 and 31.5 µm (*Crystal set I*) while the other type consists of poling periods 24.4, 24.8, 25.0, 25.3, 25.6, 26.0 and 26.4 µm (*Crystal set II*, see Sec. 3.1.1).

4.1 Primary parametric process

At pump powers slightly above the pump threshold, only one peak appears in the spectrum of the resonant waves, which can be identified as the signal wave at a wavelength of the primary parametric process λ_{s1} pumped with light of the wavelength 1030 nm (see Sec. 2.3.1). Such a process is well known to occur in cw optical parametric oscillators [33]. Its tuning behaviour for our particular system is characterised in this section. The signal wavelengths are measured directly with the wavemeter while the corresponding idler wavelengths are calculated via the resonance condition (Eq. (2.2)).

Figure 4.1 shows these signal and idler wavelengths of the primary oscillation with respect to the crystal temperature, which is varied between 40 and 160 °C. The left-hand side of the graph illustrates the tuning for *Crystal set I* while the right side shows the tuning for *Crystal set II* poling structures. The smallest signal wavelengths on the right correspond to a QPM period of 28.5 µm and the largest signal wavelengths to 31.0 µm. For a QPM period of 31.5 µm no phase matching is possible at all, which is why no data could be measured. On the left, the lowest signal wavelength is generated by $\Lambda = 24.4$ µm.

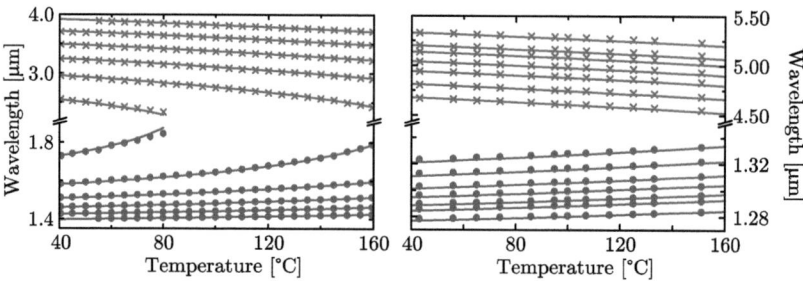

Figure 4.1: *Wavelength of the signal (red) and idler (grey) waves for different poling period lengths with respect to the crystal temperature. The symbols show the measured data points and the solid lines illustrate the theoretical tuning curves calculated from the temperature dependent Sellmeier equation of Gayer et al. [46].*

The value of λ_{s1} depends on the crystal structuring as well as the crystal temperature. In lithium niobate for the standard parametric process with a pump wavelength at 1030 nm, higher temperatures also correspond to higher signal wavelengths due to the dispersion. It can be seen in Fig. 4.1 that theory and measurement agree nicely. Continuous tuning is achieved between 1.27 and 1.34 µm and also between 1.40 and 1.84 µm for the signal waves. Here, *continuous* means that every wavelength within these regions can be addressed. The corresponding idler waves are 4.55 to 5.32 µm and 2.33 to 3.89 µm. The highest idler wavelength is generated with the smallest QPM period of 24.4 µm. For a mode-hop-free change of the wavelength, the piezo element at the plane mirror can be moved, enabling 1 GHz of fine tuning.

The tuning performance of our IR OPO is so far limited by the poling periods available and the crystal coatings. Closer to degeneracy at 2.06 µm, single-frequency operation becomes more challenging due to broader parametric gain profiles in our type 0 phase matching [71], but has been demonstrated in similar systems already [72]. In this thesis, we avoided going to degeneracy since here the OPO becomes doubly-resonant and for such systems our theories do no longer apply (see Sec. 2.2.2).

Towards higher wavelengths above 5.3 µm, the absorption caused by the damped phonon resonance at 16 µm (18.8 THz) starts to impact OPO operation [73]. An additional resonance is present at 14.5 µm (20.8 THz). This transition would be closer to our idler wavelengths, but the strength of this resonance ist smaller, reducing its influence.

In literature, tuning of continuous-wave optical parametric oscillators based on lithium niobate between 2.7 and 4.8 µm of idler wavelengths is reported [36, 74], which is smaller than the range covered by our device. For signal wavelengths around 1550 nm the pump thresholds are typically below 1 W (see Sec. 3) and the idler output can be as high as 3.5 W. At 5.3 µm the output powers that we achieve are in the order of 1 mW only and the pump threshold is 3.5 W. For larger wavelengths, the growing absorption will increase the threshold such that the maximum pump power of 19 W will no longer be sufficient to start the oscillation. However, our value of 5.3 µm is the highest generated in such devices based on lithium niobate which has been reported so far.

4.2 Spectral characteristics for processes involving terahertz waves

4.2.1 Spectral features at different pump powers

In the section above, only low pump levels with one resonant component being visible in the spectrum have been addressed. Figure 4.2 shows spectra taken with the Agilent spectrometer at four different pump powers. The pump threshold P_{th} for the first resonant peak to appear is 0.9 W. Slightly above this threshold, at a pump power of 1 W, still only one peak is visible in the spectrum at a wavelength of λ_{s1}, being the primary parametric process as described above. Additional components occur if the pump power is increased. At twice the pump threshold, two peaks can be seen in the spectrum. The number of resonant components increases further with rising pump powers until numerous wavelength components can be seen at $P_p = 9$ W, which can be grouped into two categories: On the one hand, the primary peak seems to broaden, while on the other hand, new components appear at characteristic frequency differences with respect to the first wavelength maximum. The broadband coatings of mirrors and crystals allow a many wavelength components to be resonant at once.

The following subsections discuss the different origins of these spectral components and their relevance for terahertz generation.

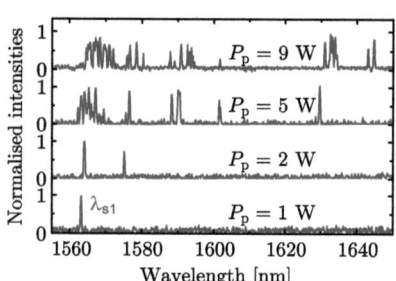

Figure 4.2: Spectra of the resonant waves at four different pump powers. For a pump power of $P_p = 1$ W only one peak at a wavelength λ_{s1} is visible in the spectrum. For higher pump powers P_p additional resonant components appear in the spectra. Experimental parameters: crystal length $L = 5$ cm, poling period $\Lambda = 30.0$ μm, crystal temperature $T = 128\ °C$.

Parametric gain

The calculated gain is plotted in Fig. 4.3. All theoretical parameters needed in the calculation (see Eq. (2.12)) are chosen accordingly to match the experimental parameters corresponding to Fig. 4.2. The refractive indices needed are once more determined via the Sellmeier equation for the infrared [46].

Losses within the resonator due to imperfect anti-reflection coatings and mirror reflectivities amount to approximately 1 %. Therefore, the amplification of a light wave of a

SPECTRAL FEATURES OF THE RESONANT WAVES

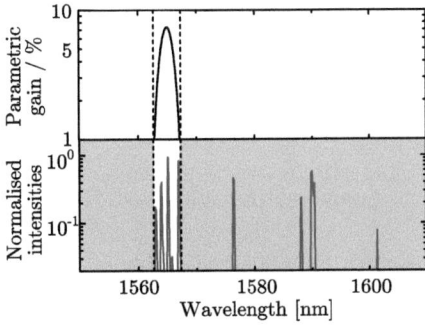

Figure 4.3: *Top: Calculated gain profile for 5 cm crystal length and a pump power of 5 W. Losses within the cavity are approximately 1 %. Only light beams with wavelengths above this threshold can be amplified. Bottom: Corresponding measured spectrum taken at $P_\mathrm{p} = 5$ W, experimental details see Fig. 4.2. Peaks in the grey region do not belong to the primary OPO process.*

certain wavelength needs to overcome this boundary in order to oscillate. If the pump power is increased, a broader wavelength range fulfills this threshold condition. This can lead to multi-mode operation of the OPO [75]. Such an explanation is simplified, since different modes are coupled, because they are fed by the same pump wave, and pump depletion occurs. Nevertheless, all accessible wavelengths need to lie within the parametric gain profile, which thus defines boundaries for the oscillation process. The calculated gain profile is plotted only to a value of 1 % amplification, since this is the value of losses to be overcome.

It can be clearly seen that just a couple of the resonant components, around a centre wavelength of 1565 nm, lie within the boundaries of parametric amplification (see Fig. 4.3, grey areas). Only these waves can be generated by the primary parametric process with a pump wavelength of 1030 nm. This broadening of the primary process is not desired, since we intend to generate single-frequency terahertz radiation. Therefore, the conclusions from the previous chapter will be employed in the following chapter, using only pump powers below the Kreuzer limit. For single-mode operation, in Chapt. 3 we have stopped recording the powers when additional components appeared within this parametric gain. The origin of the other remaining resonant wavelengths outside the gain profile will no be investigated.

4.2.2 Tuning behaviour of difference frequencies

So far, we have only studied the tuning of the first resonant peak (see Sec. 4.1). Now, we want to concentrate on the frequency differences of the other resonant components with respect to the primary signal wavelength λ_s1. For this, we take a closer look at these frequency separations in a spectrum with only one peak within the parametric gain, as exemplarily illustrated in Fig. 4.4. The absolute value of the signal wavelength is shifted because the QPM period used is $\Lambda = 25.6$ μm instead of $\Lambda = 30.0$ μm (see Sec. 4.1). In this graph four wavelength components in addition to the first peak λ_s1 can be seen.

SPECTRAL FEATURES OF THE RESONANT WAVES

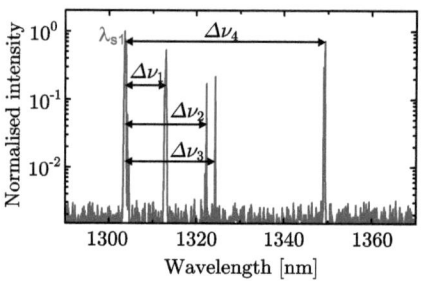

Figure 4.4: *Spectrum of the resonant waves taken by the Agilent spectrometer. Four spectral components can be seen, the first labeled with λ_{s1}. The frequency separation between λ_{s1} and the other components are denoted by $\Delta\nu_1$, $\Delta\nu_2$, $\Delta\nu_3$ and $\Delta\nu_4$ with values of 1.6, 3.2, 3.6 and 7.6 THz, respectively. Experimental parameters: $\Lambda = 25.6$ µm, $T = 60\,°C$ and $P_p = 8$ W.*

Their frequency separations with respect to λ_{s1} are labeled with $\Delta\nu_1$, $\Delta\nu_2$, $\Delta\nu_3$ and $\Delta\nu_4$ according to their amplitude with $\Delta\nu_1$ being the smallest difference frequency at 1.6 THz. The separation $\Delta\nu_2$ is twice as large, 3.2 THz, while $\Delta\nu_3$ amounts to 3.6 THz. The spectrum in Fig. 4.4 also shows a wavelength component at an even larger frequency separation of $\Delta\nu_4 = 7.6$ THz.

Keeping the crystal temperature fixed at 60 °C and only varying the poling period, leads to the tuning behaviour shown in Fig. 4.5. Here, the triangular symbols represent the frequency separation between the first two spectral components $\Delta\nu_1$ with a tuning capability ranging from 1.3 to 1.7 THz. Exactly twice these values can be found in the tuning of the frequency separation $\Delta\nu_2$. In Fig. 4.4, a further spectral component at a distance of $\Delta\nu_3$ is shown. Its frequency separation from the first signal wave can also be tuned. This is illustrated in Fig. 4.5 by the blue dots. The resulting frequency tuning ranges from 3.1 to 3.6 THz. In addition, the dependence of the difference frequency $\Delta\nu_4$ around 7.6 THz is plotted (open squares).

One can clearly see, that the behaviour of the various difference frequencies with respect to the poling structure is twofold. The frequency separations $\Delta\nu_1$, $\Delta\nu_2$ as well as $\Delta\nu_3$

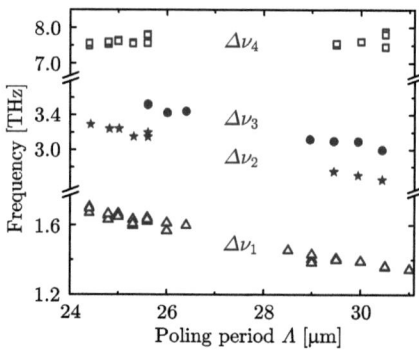

Figure 4.5: *Generated frequency separations with respect to the QPM period Λ. The open triangles represent the tuning of the frequency difference $\Delta\nu_1$ between λ_{s1} and the second resonant component while the stars stand for the tuning of $\Delta\nu_2$, which is always twice as large as $\Delta\nu_1$. The dots symbolise the separation $\Delta\nu_3$ (see Fig. 4.4). In addition, the open squares show the tuning behaviour of $\Delta\nu_4$. The crystal temperature is fixed at 60 °C.*

increase with decreasing poling period length. However, the difference $\Delta\nu_4$ remains almost constant. It varies around a value of 7.6 THz, but no dependence on Λ can be deduced. This suggests that $\Delta\nu_4$ originates from a fundamentally different process which will be discussed below.

Raman scattering

Phonon resonances present in lithium niobate crystals can be used to explain absorption effects. However, such resonances in crystals can have additional influences on incoming light waves. A well studied phenomenon is the Raman scattering effect [76], where the frequency of an incoming light wave can be shifted by the resonance frequency of the lattice vibration, i.e. the phonon. A process like Raman scattering involves no dependence on the poling period of the crystal, suggesting that $\Delta\nu_4$ might be caused by such a Raman transition. In lithium niobate, a prominent phonon resonance between 7 and 8 THz exists [77]. This is very close to the measured frequency shift $\Delta\nu_4$. When doped with magnesium oxide, the centre frequencies of the lattice vibration modes in the crystals can change slightly [78, 79]. That is why, the measured data is compared with the phonon resonance at 7.67 THz [80] for extra-ordinarily polarised light and magnesium-doped lithium niobate. Figure 4.6 shows the theoretical Raman frequency shifts in comparison with the measured values of $\Delta\nu_4$.

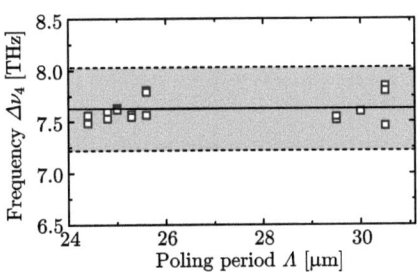

Figure 4.6: *Measured frequency difference $\Delta\nu_4$ (open squares). The solid line signifies the centre frequency of the theoretical Raman transition in magnesium-doped lithium niobate while the shaded region, restricted by two dashed lines, shows the predicted width at room temperature [81].*

Generally phonon resonances are very broad, i.e. in the order of THz. The solid line shows the centre frequency of the theoretical phonon transition at 256 cm^{-1} (corresponding to 7.67 THz) [80], while the shaded region, restricted by two dashed lines, represents the full-width-at-half-maximum FWHM of ± 26 cm^{-1} (corresponding to ± 0.8 THz) of this Raman peak [81]. All data points lie within this region, confirming the identification of this spectral component as a phonon transition. Shifts caused by phonons do not lead to the direct emission of terahertz waves and thus cannot be used to drive a terahertz OPO. Nevertheless, one could employ these two waves as a pump source for difference frequency but this would not be tunable because the Raman shifts are fixed for each nonlinear medium.

Parametric processes generating terahertz waves

So far, the origin of the broadening of the primary signal peak as well the spectral component shifted by 7.6 THz have been explained. Now we want to take a closer look at the other resonant components. In recent literature, frequency shifts $\Delta\nu_1$, $\Delta\nu_2$ and $\Delta\nu_3$ were also assigned to Raman shifts, e.g. by Okishev et al. in 2006 [82] or Henderson et al. in 2007 [83]. However, at frequencies of approximately 1.5 or 3 THz no phonon vibrations are known in magnesium-doped lithium niobate [80]. In this work, we have also shown that these shifts depend on the QPM period, ruling out that they originate from Raman scattering.

To check whether a cascaded process (see Sec. 2.4) is responsible for the extra resonant components, we compare the dependence of $\Delta\nu_1$ and $\Delta\nu_3$ on Λ with theoretical expectations for parametric processes pumped with the primary signal wave λ_{s1}. This is illustrated in Fig. 4.7. The theoretical curves are obtained considering a pump wavelength of $\lambda_{s1} = 1550$ nm for the refractive indices used [46, 47]. All measurements are made at 60 °C, therefore the room temperature Sellmeier coefficients from [47] can still be used for obtaining reasonably good results.

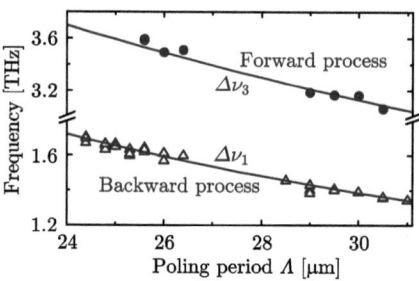

Figure 4.7: *Tuning behaviour of the difference frequencies by a cascaded forward or backward parametric process. The symbols show measured data points. Solid lines represent the theory according to the infrared [46] and terahertz Sellmeier equations [47] for the two different phase matching conditions.*

The process generating idler waves at frequencies $\Delta\nu_1$ is a backward parametric process, meaning the idler wave travels anti-parallel with respect to the other participating waves (see Sec. 2.4). In addition to this terahertz wave, it generates a signal wave at λ_{s2}. In Fig. 4.4, $\Delta\nu_2$ is introduced as well. Since its value is always twice as large as $\Delta\nu_1$, showing exactly the same dependence on the poling period, we interpret this process as a further cascade, pumped by the secondary signal wave λ_{s2}. Several of these processes are possible within the same device and the special characteristics of such higher-order cascades are addressed in Sec. 5.3.5.

In addition to the backward process also a forward terahertz process can occur in lithium niobate within the same crystal structuring. The resulting frequency of the forward process is approximately 3 THz for QPM periods around 30 µm (see Sec. 2.3). In Fig. 4.7 the theoretical curve for $\Delta\nu_3$ corresponds to this forward terahertz process, matching the experimental data very nicely. Therefore, the spectral component corresponding to $\Delta\nu_3$

can be identified as the signal wave of a cascaded forward terahertz process. Although the oscillation threshold for this process is higher due to larger absorption it can apparently be overcome.

In the following chapter, direct verification of generated terahertz waves is presented.

CHAPTER 5

TERAHERTZ WAVE GENERATION

This chapter shows the characterisation of the first continuous-wave optical parametric oscillator. To begin with, the needed experimental methods needed in addition to the OPO setup for infrared waves are presented. In the following, the experimental realisation of the THz OPO based on cascaded parametric processes is demonstrated. Different crystal properties in the infrared and the THz regime are compared and the performance of the terahertz OPO in comparison with other methods of THz generation is discussed.

5.1 Experimental methods for the terahertz setup

Based on the analysis from Chapt. 3, a standard OPO with a 2.5 cm long crystal is used for terahertz generation as a compromise between high powers and low thresholds. Since high signal powers are essential, only highly-reflecting cavity mirrors are taken.

5.1.1 Terahertz wave detection

To extract the backwards traveling terahertz wave from the cavity (see Sec. 2.4), we place an off-axis-parabolic mirror into the resonator such that its distance to the crystal centre is approximately equal to its focal length of 2.5 cm. This mirror should then collimate the divergent THz beam. A hole of 1.4 mm diameter is drilled into its aluminium surface such that the infrared waves can pass through (see Fig. 5.1). The small hole does not reduce the THz reflectivity significantly.

The parametrically generated terahertz radiation is detected with a calibrated Golay cell from the manufacturer Tydex Inc. This Golay cell is a thermal detector, in which the curving of a metal-coated membrane, connected to a gas cell, is measured [84]. A maximum power of 10 µW can be taken by the detector before destroying the fragile membrane. The calibration is 80 kV/W, as specified by the manufacturer, if the incoming radiation is chopped with 10 Hz. The noise-equivalent-power is stated to be 100 pW/$\sqrt{\text{Hz}}$.

To ensure that as much terahertz radiation as possible actually passes through the incidence window, the collimated terahertz beam is focussed onto the Golay cell with an additional off-axis parabolic mirror PM2 with a focal length of 5 cm (see Fig. 5.1). We call this experimental setup *THz Setup I*. A lock-in amplifier processes the registered signals. The incidence window consists of diamond with a 5 mm aperture and a transparency range of 0.25 to 4 µm and 6 to 8000 µm. Due to the large transmission range of the Golay cell entrance window, one has to prevent residual visible and infrared radiation from entering. Therefore, a filter made of blackened high-density polyethylene HDPE made by GSE Lining Technology GmbH with a thickness of 1 mm is placed directly in front of the Golay cell. The transmission of this filter at 1.5 THz is measured to be 25 %. All detected terahertz powers are corrected according to this attenuation.

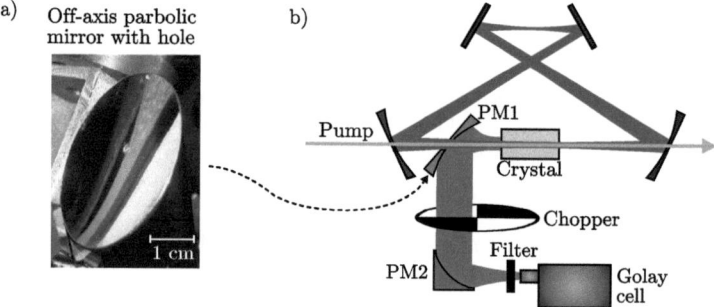

Figure 5.1: a) Off axis parabolic mirror with a hole in its middle. b) Schematic setup for detecting the continuous-wave terahertz power. The backwards propagating terahertz wave diverges when leaving the crystal and is collimated by the first parabolic mirror PM1. A second parabolic mirror PM2 (focal length 5 cm) focusses the terahertz light onto a Golay cell. A filter in front of this detector strongly attenuates the infrared radiation. Between PM1 and PM2 the THz wave is chopped.

To estimate whether much terahertz radiation is lost at the hole in the mirror, the divergence of the terahertz beam needs to be calculated using ABCD-matrix formalism [67,85]. The resulting $1/e^2$-radius for the terahertz wave is displayed in Fig. 5.2. Due to its larger wavelength (around 200 µm), we assume that the terahertz beam fills the entire crystal height of 0.5 mm at the end surface and therefore we start the calculation with a beam radius of $w = 0.25$ mm. At the position of the mirror the terahertz beam has already got a diameter of about 1.4 cm (see Fig. 5.2) and thus covers most of the mirror surface. Therefore, it should only loose a small amount of its power at the hole in the mirror.

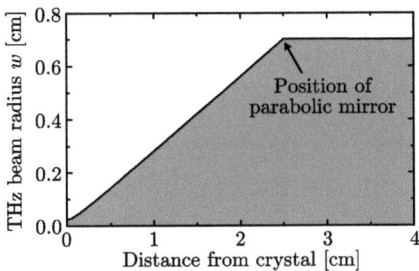

Figure 5.2: *Radius w of the terahertz beam with respect to the distance from the crystal end facet. The THz wave is assumed to leave the crystal with $w = 0.25$ mm, corresponding to half the height of the crystal itself. At a distance of 2.5 cm to the crystal surface we place a parabolic mirror with a focal length of 2.5 cm, which collimates the divergent terahertz beam.*

5.1.2 Measuring the terahertz beam polarisation

The wavelength of terahertz radiation is by two orders of magnitude larger than those of the participating infrared waves. This feature makes the fabrication of optical components easier since less precision is needed and polishing becomes simple. A THz polariser can thus be built as a parallel grid of conducting wires [86] and inserted into the setup at a position, where it intersects a collimated terahertz beam as shown in Fig. 5.3. Behind the polariser, the terahertz beam is then focussed onto the Golay cell. All infrared waves are extra-ordinarily polarised with respect to the optical axis of the lithium niobate crystal. In the employed phase matching scheme the terahertz wave has linear polarisation as well.

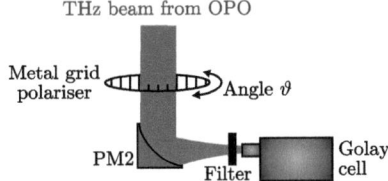

Figure 5.3: *Experimental setup of a metal grid polariser placed in a collimated terahertz beam between parabolic mirrors PM1 and PM2, generated within a nonlinear crystal. The polariser can be turned by the angles ϑ to analyse the influence of the grids onto the terahertz power.*

We produce such terahertz polarisers by winding gold-coated tungsten wires around metal frames with the desired spacing. The optimum diameter of these wires is less than $\lambda_{THz}/10$ and their spacing less than $\lambda_{THz}/4$ [87]. The parallel component of an incident THz field will be reflected almost completely, since the conductive circuit is closed. For the perpendicular component the dipole constituents cannot oscillate and therefore the field can pass the polariser. The width of the wires used is $a = 30$ μm, a size that is already close to $\lambda_{THz}/10$ for $\lambda_{THz} = 220$ μm, corresponding to 1.4 THz. The wire spacing for a metal-grid polariser is chosen to be $b = 60$ μm (see Fig. 5.4a), which is close to one fourth of the incident wavelength λ_{THz}. Putting this metal grid polariser into the collimated terahertz beam of *THz Setup I* and turning the polariser should provide a sinusoidal modulation of the detected power as illustrated in Fig. 5.4b.

Terahertz wave generation

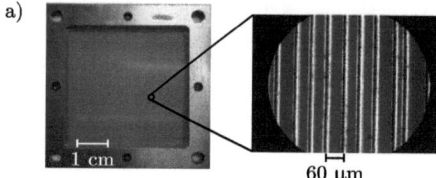 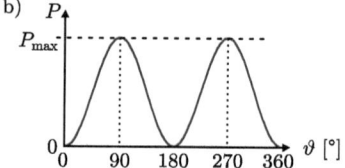

Figure 5.4: a) Metal grid polariser for terahertz radiation made of gold-coated tungsten wires with a width of 30 µm and a spacing of 60 µm. b) Theoretically expected behaviour of the detected terahertz power, when turning the polariser grid.

The influence of the metal grid onto pump, signal and idler powers is shown in Fig. 5.5. Due to the size and separation of the wires, the polariser acts only as a shadow mask for the pump, signal and idler waves. The infrared waves, if part of them still reaches the detector, can thus only add a constant background to the measurement.

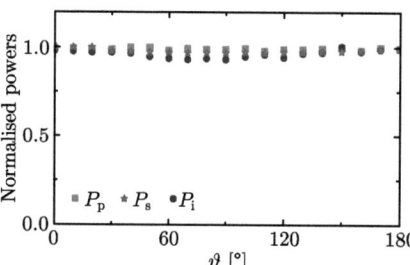

Figure 5.5: Measured normalised pump P_p, signal P_s and idler powers P_i with respect to the angle ϑ of the terahertz polariser. Here, the orientation of the metal grid is defined as zero when the wires are parallel to the polarisation of the infrared beams. No increase in any of the three powers at 90° can be seen as it is expected for terahertz waves (see Fig. 5.4b).

5.1.3 Determining the terahertz wavelength and linewidth

Besides polarisation properties, a further proof for the existence of terahertz waves is measuring the wavelength of the light hitting the detector. This can be done by a Fabry-Pérot interferometer (FPI) which is placed into the collimated terahertz beam instead of the polariser as depicted in Fig. 5.6. It consists of two mirrors, one on a translation stage, such that the distance D between the mirrors can be varied automatically to realise a scanning FPI. The translation stage is attached to a computer-driven stepper motor. The total length of the FPI is approximately 1 cm.

Similar to the grid polarisers, we can construct mirrors for a terahertz FPI ourselves. If the wires are crossbred, they form a mesh, resulting in partly transmitting mirrors for terahertz waves (see Fig. 5.7a). This time we employ aluminium wires with a diameter of $a = 15$ µm. For a mirror with a measured transmission of 70 % the spacing is $b = 100$ µm. If one mirror is moved, varying D, the detected terahertz power should show the behaviour

Terahertz wave generation

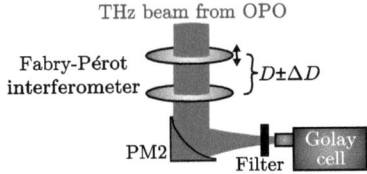

Figure 5.6: *Schematic setup of a terahertz Fabry-Pérot interferometer (FPI) positioned in the collimated terahertz beam between the two parabolic mirrors PM1 and PM2. The distance D between the two mirrors of the FPI can be varied by ΔD.*

displayed in Fig. 5.7b, experiencing peaks. Two maxima in power are separated by half the wavelength of the incident light [56]. The metal meshes are partly transparent, making a pre-adjustment with visible radiation possible.

This terahertz FPI can in general also be used to determine the spectral linewidth of the THz beam. For a total length of $D = 1$ cm, the free spectral range FSR is 15 GHz. Due to the low finesse of only 4, the smallest linewidth that is possible to measure is 3.75 GHz. A more precise estimate of the THz linewidth can be achieved via measuring the linewidth of the signal waves with the FPI for infrared waves (see Sec. 3.1.2). Since the terahertz wave is the difference frequency of the two signal waves λ_{s1} and λ_{s2}, its linewidth should be approximately the sum of both signal linewidths from primary and secondary process.

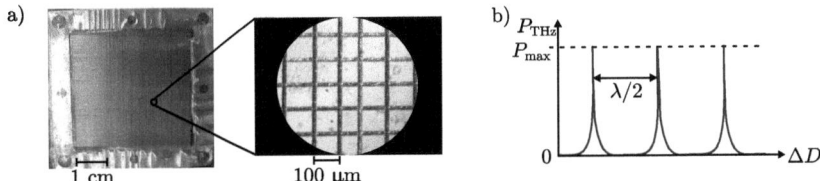

Figure 5.7: *a) Partly transparent mirror for terahertz radiation formed by an aluminium wire mesh. The distance between the wires is $b = 100$ µm and their width $a = 15$ µm. b) Theoretically expected behaviour of the detected terahertz power, when varying the length of the Fabry-Pérot interferometer by ΔD. The spacing between to peaks in the detected terahertz power is half the wavelength λ of the incident light.*

5.1.4 Characterising the terahertz beam dimensions

For most applications it is important to have a small focal spot with a circular shape. To determine the terahertz beam dimensions in the focus, it is useful to create two focal spots of terahertz radiation: one that is evaluated and the other one at the position of the Golay cell. For this purpose, the setup can be extended by two more off-axis parabolic mirrors PM3 and PM4 with focal lengths of 5 cm. This setup configuration is called *THz Setup II* (see Fig. 5.8).

Terahertz wave generation

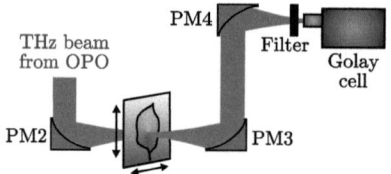

Figure 5.8: *The terahertz detection setup is extended by additional parabolic mirrors PM3 and PM4, giving two focal spots in the THz beam. This way a sample can be scanned through the first focus while the Golay cell detects the light at the position of the second focus.*

To determine the beam size and its elipticity, we use a special sample, a pie chart made of aluminium foil glued to a piece of polystyrene (see Fig. 5.9). Due to the pointed openings, one can observe the resolution in the middle of the sample. The polystyrene is only 1.8 mm thick and its transmission is about 90 % at 1 THz. Besides the pie chart, other samples can be scanned through the terahertz beam to analyse their absorption properties. The motorised translation stages can cover a distance of 5 cm in each direction with a possible resolution of 2.5 µm. However, the resolution for scanning a sample is limited by the focus size of the terahertz beam.

Figure 5.9: *Pie chart to measure the beam size of the terahertz wave as well as its elipticity. A piece of aluminium foil with triangularly shaped openings is glued to polystyrene of 1.8 mm thickness. The polystyrene is almost transparent for terahertz radiation whereas the aluminium foil, as a metal, reflects the terahertz wave. The rectangular hole at the lower edge is used to calibrate the dimensions and orientation.*

In addition to this technique, we use a knife edge, attached to a two-dimensional translation stage. The knife edge is scanned vertically or horizontally through the THz beam, while the intensity is measured by the Golay cell. A Gaussian beam shape is fitted to the resulting intensity profile. Such a measurement at different distances to the focussing mirror an analysis of the beam quality can be performed [88].

Having introduced all these methods, one can now take a look at the characteristics of the experimentally realised cw terahertz OPO.

5.2 Terahertz optical parametric oscillator

This section contains measurements, characterising the performance of our THz OPO. Figure 5.10 shows a picture of the cavity including the additional off-axis parabolic mirror PM1. This mirror reflects the terahertz wave out of the resonator, but transmits pump and signal beams through a hole in its centre. In contrast to the photograph of the IR setup (Fig. 3.4), here a smaller oven can be seen because the nonlinear crystal is shorter.

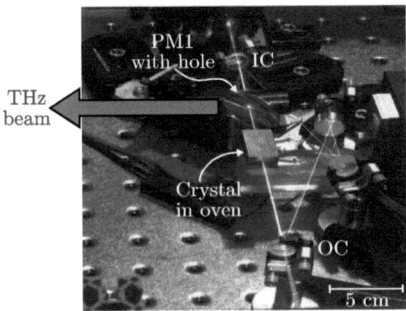

Figure 5.10: *Photograph of the terahertz optical parametric oscillator. The blue-purple beam is the 1030 nm pump wave, while the green line shows the path of the resonant signal waves. IC and OC label the incoupling and outcoupling curved cavity mirrors as introduced in Sec. 3.1. The off-axis parabolic mirror with a hole in its centre reflects the terahertz wave out of the resonator towards the Golay cell for detection.*

The infrared pump beam at 1030 nm, going through the two curved mirrors IC and OC, is seen by the camera (Fig. 5.10, blue-purple line). The path of the signal waves within the resonator is once more visualised by frequency-doubled light (Fig. 5.10, green line) and sum frequencies, processes that are not phase matched. The terahertz wave cannot be photographed and is therefore sketched in the picture.

In the following sections, the properties of the terahertz wave such as power, polarisation, linewidth, beam shape and stability are measured. The thresholds needed to start terahertz processes are evaluated for different crystal lengths. Furthermore, the frequency tuning characteristics are presented and scan applications demonstrated. Additionally, the influence of higher-order cascaded processes on the terahertz output is shown.

5.2.1 Power measurements

The terahertz beam, leaving the resonator, is focussed into a Golay cell (see Sec. 5.1, *THz Setup I*). Power measurements of the THz wave, as displayed in Fig. 5.11, are performed with a 2.5 cm long crystal at $\Lambda = 30.0$ µm and 120 °C and highly-reflecting cavity mirrors, i.e. no outcoupling mirror is used for the signal light and hence the intra-cavity signal power maximised. Terahertz power should be present only if the second signal wave is present in the spectrum (see Sec. 2.4). Therefore, we take spectra of the leaking signal fields and observe simultaneously the power entering the Golay cell.

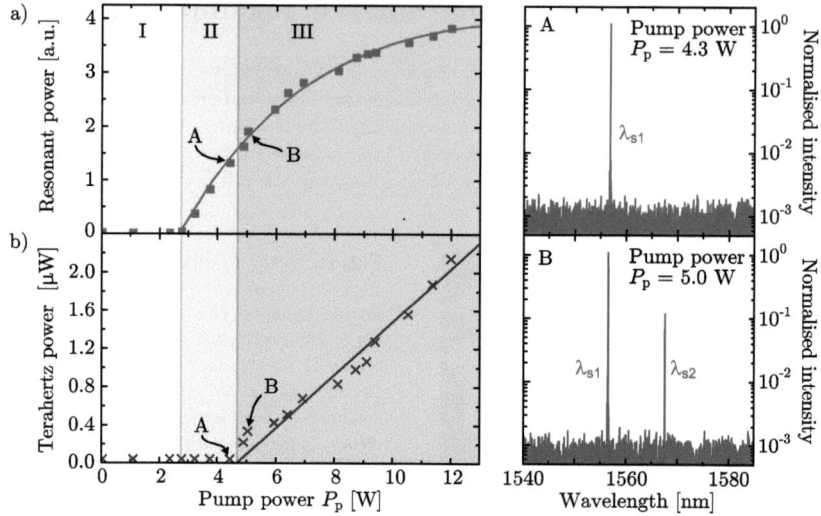

Figure 5.11: *Left side: a) Resonant power with respect to the initial pump power P_p. b) Terahertz power measured with the Golay cell with respect to P_p. Solid lines are just guides to the eye. Three coloured regions indicate the presence of no parametric process (I), only primary process (II) and additionally a terahertz-generating parametric process (III). Right side: spectra of the resonant waves. At $P_p = 4.3$ W only one peak is seen in the spectrum (Point A). This peak is marked with λ_{s1}, the wavelength of the primary signal wave. Point B shows the spectrum at 5.0 W pump power, in which two signal waves, λ_{s1} and λ_{s2} from the first and second parametric process, can be seen.*

The terahertz wave is generated within the cavity by the high-power signal wave of the primary parametric process. In Chapt. 4, it is indicated that two thresholds should be visible in the power measurements, one for the primary and the other one for the terahertz process. Figure 5.11 shows the behaviour of the resonant signal power and the corresponding terahertz output power with respect to the initial pump power P_p of the primary pump wave at a wavelength of 1030 nm. Both, the resonant power as well as the THz power, rise with increasing pump powers once their threshold is reached, but experience different slopes.

In general, three different regions (I, II, III) can be distinguished. The white region I indicates that no parametric process is present and thus no peak can be seen in the spectrum. In the yellow-coloured area II, starting at $P_p = 2.8$ W, only the signal wave of the first parametric process is present. At $P_p = 4.3$ W a spectrum, showing only one peak, can be seen (Point A, Fig. 5.11, right side). The third, orange-coloured region III, which starts at a primary pump power of $P_p = 4.7$ W, marks the presence of the second parametric process, actually generating the terahertz wave. This is indicated by an increasing power as detected by the Golay cell and the additional component in the resonant spectrum (see Fig. 5.11, right side, spectrum at point B, taken at $P_\text{p} = 5$ W). The power of the THz beam grows linearly while the power of the resonant waves rises slower than linear.

Besides the primary pump values, we determine the power of the resonant wave within the cavity P_s1 at the onset of the cascaded process, because this wave is responsible for terahertz generation. Since one cannot measure the power directly within the cavity, we detect the power outside the resonator behind a plane mirror with 0.5 % residual transmission which allows to calculate the intra-cavity power.

Crystals of different lengths are available. We define the threshold of the cascaded process by the appearance of the second signal wave in the spectrum analyzer λ_s2. All signal components contribute to the resonant power and are not separated by filters. Table 5.2 shows the results of the thresholds measured for the secondary OPO process. The lowest threshold of 170 W resonant power is achieved for the longest crystal $L = 5$ cm. For a three times smaller crystal, $L = 1.7$ cm, the value of $P_\text{th,res} = 590$ W is 3.4 times higher. With high idler absorption, the threshold no longer depends quadratically on the crystal length as underlined by this measurement (see Secs. 2.1.2 and 2.2.2).

L / cm	$P_\text{th,res}$ / W
5.0	170
2.0	330
1.7	590

Table 5.1: *Threshold values of the resonant power for the terahertz parametric process $P_\text{th,res}$ with respect to the crystal length L. The threshold is defined by the appearance of the second resonant component in the wavemeter.*

5.2.2 Polarisation measurements

The diamond window of the Golay cell is transparent for all participating wavelengths. Although filters are used to cover it, it is possible that small fractions of the infrared powers reach the detector. This contribution would also rise with increasing pump power. Therefore, special attention has to be paid to proving that the power detected by the Golay cell originates from the terahertz wave.

All infrared waves are linearly polarised and according to the phase matching scheme the terahertz wave should have linear polarisation, too. In Sec. 5.1.2, self-made polariser grids for terahertz waves were introduced. Figure 5.12 shows the resulting terahertz power behind the polariser for an initial pump power of 7.4 W. All other experimental parameters are kept as above and a high-finesse cavity is used. The polariser is turned

Terahertz wave generation

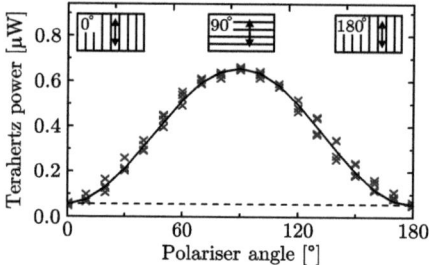

Figure 5.12: *Terahertz output power with respect to the rotation angle of the metal grid polariser (see Sec. 5.1.2). The blue crosses show the measured data points, while the solid black line is a sinusoidal fit to the data. A dashed line marks the baseline at 65 nW. The insets show the orientation of the wires in comparison with the terahertz wave polarisation.*

back and forth twice to show reproducibility. This results in four data points for each grid angle. A maximum power of 0.7 µW is reached for 90° orientation.

The behaviour of the terahertz power corresponds very well to that of linear polarisation since the data points follow the sinusoidal fit. In Fig. 5.12, the insets show the expected orientation of the terahertz wave polarisation relative to the metallic wires. Although no mode-hop occurred during the measurement time, the powers of different data points at the same polariser angle can vary by ±0.01 µW. This might be due to the error in adjusting the angle of the polariser exactly, which amounts to ±3°.

5.2.3 Linewidth

To additionally confirm that the power seen by the Golay cell actually originates from incoming terahertz radiation, we use a Fabry-Pérot interferometer measurement to determine the incident wavelength (see Sec. 5.1.3, *THz Setup II*). The distance between the two meshes, acting as partly transmitting mirrors, is varied, thus changing the total length D of the FPI. Figure 5.13 illustrates the normalised terahertz power measured behind the FPI.

The measurement shows a periodic change in the terahertz power between 100 and 10 %. The distance between two peaks is 111 µm in average, which gives a wavelength of 222 µm

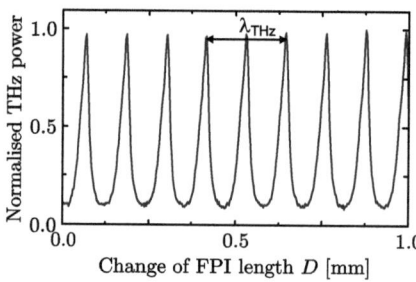

Figure 5.13: *Normalised terahertz power with respect to the change in distance D between the two mirrors of the terahertz Fabry-Pérot interferometer. A periodic change in the terahertz power can be seen. This measurement is performed for an initial pump power of $P_p = 5.2$ W. Other parameters: $L = 2.5$ cm, $\Lambda = 30.0$ µm, $T = 120\,°C$.*

agreeing exactly with the 1.35 THz determined from the spectra (see also Chapt. 4). The full-width-at-half-maximum FWHM linewidth of the resulting peaks is 3.75 GHz, corresponding to the resolution limit of the THz FPI (see Sec. 5.1.3). This value gives an upper limit of the terahertz linedwidth.

Fundamentally, the linewidth of the terahertz light depends on the linewidths of the generating waves, the signal waves of primary and cascaded parametric process, λ_{s1} and λ_{s2} (see Sec. 5.1.3). Figure 5.14 shows the measurement of the first two resonant components, λ_{s1} and λ_{s2}, as seen by the FPI for infrared waves. The free spectral range FSR is clearly visible in the graph, being 1.5 GHz. The determined linewidth FWHM, of both signal waves is 8 MHz each, which corresponds to the resolution limit of the FPI. One can see that the intensity of the second signal wave λ_{s2} is lower than the one of λ_{s1}, which is confirmed by the spectra of the resonant waves as taken by the wavemeter. The resulting linewidth of the terahertz wave according to this measurement is approximately the sum of both FWHM of the signal waves and thus 16 MHz, still providing only an upper limit.

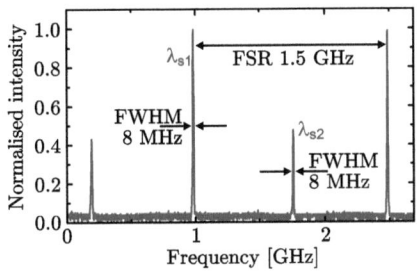

Figure 5.14: *Normalised intensity of primary and secondary signal wave, λ_{s1} and λ_{s2}, as measured by the Fabry-Pérot interferometer for infrared waves with a free spectral range FSR of 1.5 GHz. The determined full-width-at-half-maximum FWHM of λ_{s1} and λ_{s2} is 8 MHz each.*

5.2.4 Spatial beam shape

Terahertz waves can be used in imaging systems [89]. In particular the beam size is an important property (see also Sec. 5.2.8). In Chapt. 5.1, different ways to determine the size of the terahertz beam are presented (for *THz Setup II*). Figure 5.15 shows the scan of the aluminium pie chart on polystyrene (see also Fig. 5.9, original sample). The horizontal lines visible in the scan correspond to mode hops of the OPO, slightly changing the terahertz power.

Since polystyrene transmits most of the terahertz radiation, the aluminium foil can be seen as a black shadow. Here, black signifies no terahertz transmission while white corresponds to a maximum transmission of 0.2 µW terahertz power. In Fig. 5.15, the centre part of the pie chart is magnified. It can be seen, that the resolution is roughly 2 mm and the beam appears to be circular.

A knife edge can be employed as a more precise tool for measuring the beam dimensions

TERAHERTZ WAVE GENERATION

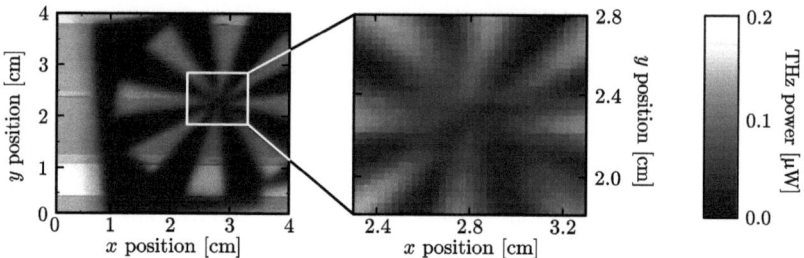

Figure 5.15: *Terahertz transmission picture at 1.35 THz of an aluminium pie chart on polystyrene. Experimental parameters: initial pump power $P_p = 10$ W, poling period $\Lambda = 30.0$ µm and crystal temperature $T = 122$ °C.*

(see Sec. 5.1.4). Similar to the pie chart it provides an estimate of the beam radius in two dimensions if vertical and horizontal scan directions are used one after the other. Figure 5.16 shows the determined beam radii at varying distances to the second off-axis parabolic mirror PM2 obtained from horizontal scans. The measured beam radius in the focal spot is 1.7 mm. A similar measurement is performed with the knife edge, being moved vertically through the beam. The smallest resulting beam radius is then 1.8 mm, confirming an almost circular shape. The measured data can be compared to the theoretical description of Gaussian beam propagation [56, 88].

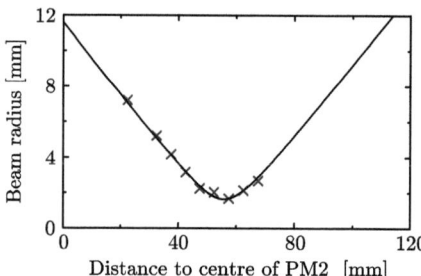

Figure 5.16: *Terahertz beam radius at different distances to the second off-axis parabolic mirror PM2. At each distance the beam radius is determined, assuming a Gaussian intensity distribution. The initial pump power is 8 W. The blue crosses are the measured data points, while the black solid line shows a fit to the data according to [56, 88].*

5.2.5 Stability

Usually, the two waves of primary and secondary process with wavelengths λ_{s1} and λ_{s2} show the same behaviour. If λ_{s1} experience a jump, λ_{s2} will change as well. However, also changes in only one of the two resonant wavelengths can be observed as displayed in Fig. 5.17 without active stabilisation. Here, a frequency jump of 2.8 GHz occurs for the wave λ_{s2} while the other remains mode hop free.

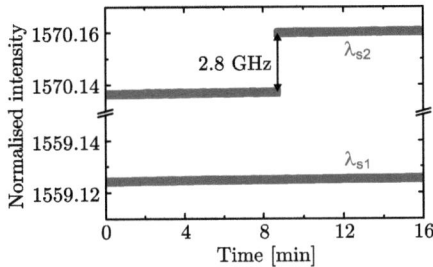

Figure 5.17: Wavelengths λ_{s1} and λ_{s2} of the resonant waves vs. time. While λ_{s1} remains constant, a jump of 2.8 GHz occurs in λ_{s2}. Experimental parameters: crystal length $L = 2.5$ cm, QPM period $\Lambda = 30.0$ µm, crystal temperature 125 °C.

In the scan of the pie chart one can see that mode hops should be avoided, because they can create unwanted patterns within the transmission pictures. Ideally, frequency of the terahertz wave as well as its power should not vary significantly during the measurement time. This can be achieved by using the active stabilisation introduced in Sec. 3.1.2. Figure 5.18 shows the detected THz power observed over one hour while a scan is performed. During this time no mode hops in the wavelengths occur in the recording of the resonant wavelength components implying a constant terahertz frequency. The power fluctuations of the terahertz wave are ±5 %.

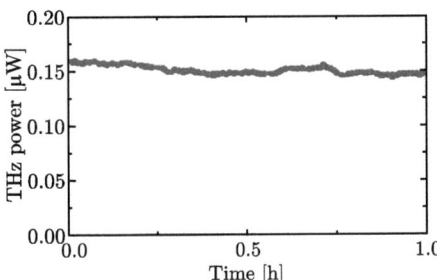

Figure 5.18: Measured terahertz power as detected by the Golay cell over one hour. Experimental parameters: $P_p = 10$ W, $\Lambda = 30.0$ µm and $T = 122$ °C in THz setup II.

5.2.6 Tuning of parametrically generated terahertz waves

Optical parametric oscillators are widely used in spectroscopy because of their large tuning ranges. In Chapt. 4, the tuning of the infrared waves is discussed. Here we want to take a closer look at the tuning behaviour of our system in the terahertz range.

We have already seen that a change of the QPM period alters the THz frequency (see Fig. 4.7). Now, the terahertz tuning with respect to the crystal temperature is investigated, as shown in Fig. 5.19 for five different poling periods (28.5, 29.0, 29.5, 30.0 and 30.5 µm). The terahertz frequency is determined by the frequency separation of the first two resonant components with the wavemeter and confirmed by a terahertz FPI measurement. For each poling period, the temperature is varied between 50 and 160 °C. The resulting tuning is 1.32 to 1.45 THz. Here, only the backward parametric process is considered, since this has been confirmed by power measurements (see Sec. 5.2.1).

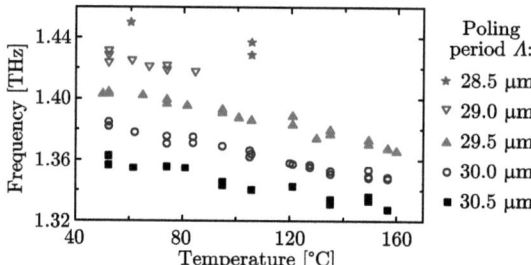

Figure 5.19: *Tuning behaviour of the parametrically generated terahertz waves with respect to the crystal temperature for different poling periods Λ. The resulting frequencies range from 1.32 to 1.45 THz.*

5.2.7 Terahertz output power in relation to higher-order cascades

In Sec. 4.2.2 it is already shown that more than one cascaded backward process is possible (see Fig. 4.4). This effect will now be investigated in more detail. For high initial pump powers, the number of peaks in the spectrum, recorded by the wavemeter, increases until up to seven resonant components are visible.

The frequency separations of all resonant wavelength components can be pursued over time with the wavemeter as displayed in Fig. 5.21, left side. Determining the difference frequency between two neighbouring components from this measurement leads to the result shown in Fig. 5.21, right side, giving a difference frequency of 1.35 THz each. All difference frequencies are equidistant within the accuracy of the wavemeter. The value of this frequency spacing agrees with the one measured by the THz FPI (see Sec. 5.2.3).

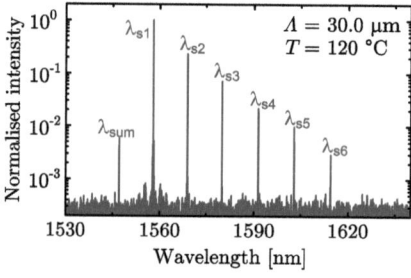

Figure 5.20: *Spectra of the resonant waves for a pump power of $P_p = 11$ W. For high initial pump powers the number of peaks in the spectrum increases. The different components are labeled with λ_{sl}, l ranging from 1 to 6. In addition, light of a lower wavelength than λ_{s1} occurs, marked with λ_{sum}.*

TERAHERTZ WAVE GENERATION

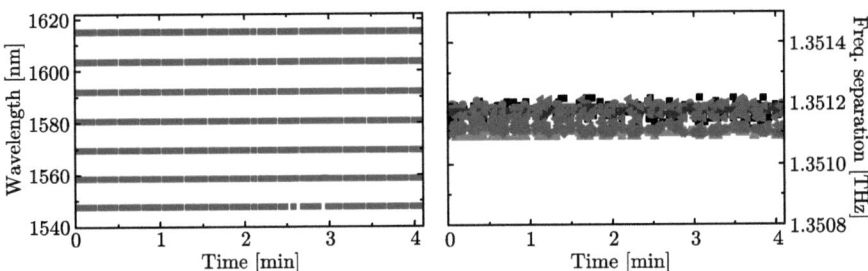

Figure 5.21: *Left side: Wavelengths of the resonant components vs. time. The measurement is performed at a pump power of $P_p = 11$ W, a poling period length of $\Lambda = 30.0$ µm and a crystal temperature of $T = 120$ °C. Right side: Difference frequency of all directly neighbouring components from the measurement on the left side, represented by various symbols. This difference frequency is approximately 1.351 THz.*

Besides wavelength and frequency difference measurements, further investigations can be performed with these additional resonant components. In the following, their influence on the terahertz power and on the pump depletion is presented. In Fig. 5.17 we have already seen that mode hops can occur in the secondary signal wave while the primary one remains constant, showing that the cascaded process is more sensitive to cavity drift. Thus, the piezoelectrically supported mirror mount should be able to control the number of wavelength peaks in the spectra. If one varies the voltage applied to the piezoelectric element, the number of resonant components changes, as exemplified in Fig. 5.22 (left side).

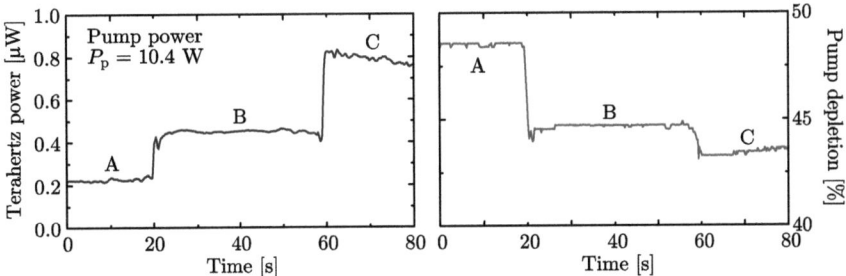

Figure 5.22: *Left side: terahertz output power vs. time. Right side: simultaneously recorded pump depletion. The points A, B and C label regions, where respectively three, four and six resonant wavelength components are present. Both measurements are performed at $\Lambda = 30.0$ µm and $T = 120$ °C as before.*

63

TERAHERTZ WAVE GENERATION

At a terahertz power of 0.2 µW, three resonant components are present (Fig. 5.22, Point A). When changing the voltage applied to the piezoelectric element, a fourth component appears (Fig. 5.22, Point B) and the output power increases to 0.4 µW. A further change of the piezo element voltage leads to six resonant components and a terahertz power of 0.8 µW. The corresponding pump depletion (Fig. 5.22, right side) drops from 48 % to 43 %. This measurement is performed at an initial pump power of $P_\mathrm{p} = 10.4$ W, other parameters used are the same as above.

The number of resonant components can be related to the output terahertz power, which is presented in Fig. 5.23. Here, the power data is taken with the Golay cell after the fourth off-axis parabolic mirror (see Sec. 5.1, *THz Setup II*). This measurement is performed with the same parameters as before, but to vary the amount of spectral components, the initial pump power as well as the voltage at the piezo mirror mount are regulated.

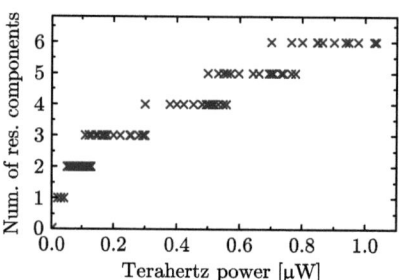

Figure 5.23: *Number of resonant components in the spectra with respect to the terahertz output power, measured with the Golay cell in the second focal spot. Other experimental parameters are the same as for the figures above but the pump power varied.*

5.2.8 Applications

Instead of the pie chart, other samples can be scanned through the terahertz focus to investigate their transmission properties. To avoid mode hops as in the pie chart scan, these scans are performed with the active stabilisation switched on as described in Sec. 3.1.2.

The step size of the translation stage is 250 µm while the integration time needed for each data point is 300 ms. Scanning a sample of 4×4 cm^2 therefore takes about an hour. In the section above, it is shown that the terahertz wave can be stable for one hour and should thus be suitable for such an application. For these measurements, the active stabilisation as described in Sec. 3.1.2 is employed.

Figure 5.24 shows exemplarily the two-dimensional transmission picture of a slightly dried lychee leave at 1.35 THz. The leave veins are clearly revealed in the terahertz picture. In this transmission scan, the lychee veins can be seen in the visible as well. However, the transmission properties of most materials are completely different in the terahertz range in comparison with those for visible or infrared radiation [90]. Therefore, we have constructed a sample out of dark-blue plastic, which is not transparent for visible light (see Fig. 5.25, left side). It is given the shape of a handbag and filled with a metallic object (a miniature gun). The terahertz transmission picture at 1.35 THz clearly reveals the content of the bag (see Fig. 5.25, right side).

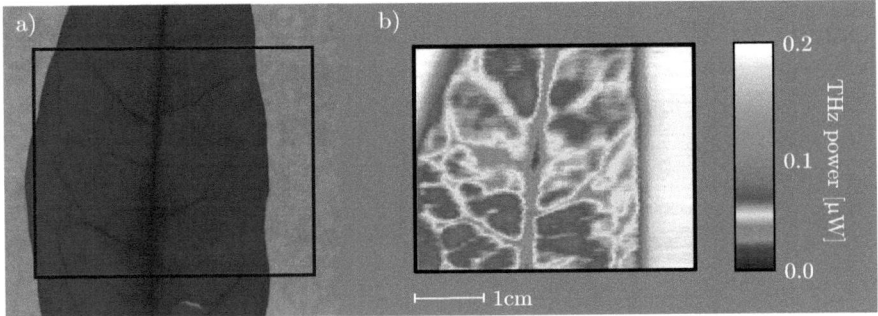

Figure 5.24: a) Photograph of a slightly dried lychee leave. b) Two-dimensional terahertz transmission of the lychee leave at 1.35 THz. White colour stands for high THz power of 0.2 µW while black signifies no terahertz transmission. Experimental parameters: initial pump power $P_\mathrm{p} = 10$ W, poling period $\Lambda = 30.0$ µm and crystal temperature $T = 122$ °C. The same terahertz power distribution of the scan as in Fig. 5.15 is shown with a different colour scale.

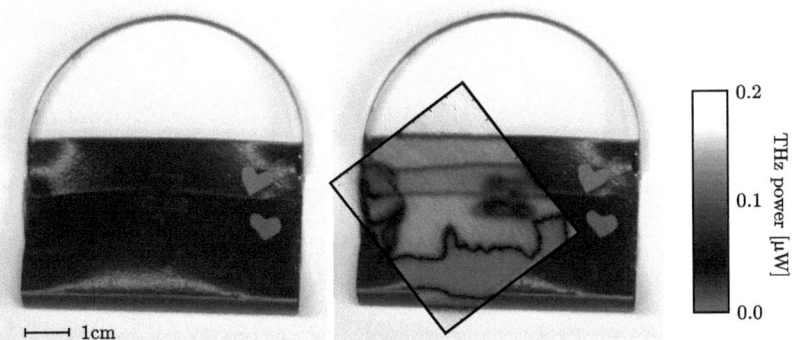

Figure 5.25: Left side: picture of a small plastic handbag not transparent for visible radiation. Right side: two-dimensional terahertz transmission scan of the plastic handbag at 1.35 THz. The red colour signifies no transmitted terahertz radiation as shown by the colour scale at the right side. A metallic gun within the bag is revealed. Experimental parameters: see Fig. 5.24.

Terahertz wave generation

In addition, other structures can be seen in the middle and at the edge of the handbag. One pattern corresponds to the fastener of the bag, while the horizontal lines are due to an overlap of plastic layers. The blob at the left edge in the transmission scan is caused by the glue used to attach different plastic pieces together. This already suggests numerous possible applications of terahertz radiation such as security scans [11], controlling medicine production [91, 92], counting layers or investigating fabrication issues [12] of terahertz transmitting materials [90].

5.3 Discussion of terahertz wave characterisation

5.3.1 Generated terahertz output power

Our experimental data unambiguously proves that indeed an OPO, generating THz waves, has been built. Now a more detailed comparison of measurements and expectations is made.

The maximum terahertz power measured in our setup is 2.2 µW. A theoretical prediction, according to Eq. (2.15), gives 10 µW, leaving a factor of 4.5 missing. Here, the power of the primary signal wave acting as a pump wave is assumed to be $P_{s1} = 400$ W while the power of the secondary one is taken to be a factor of ten smaller according to the intensity relation observed in the spectra (see Fig. 5.11). It is even possible that the signal powers within the resonator are even higher [64]. Other parameters for Eq. (2.15), e.g. wavelengths, refractive indices and crystal length, are chosen such that they match the experiment.

No anti-reflection coatings are present on the crystal surfaces for terahertz waves. Losses at the crystal surface due to reflections, amounting to almost a factor of two, are already included in the calculation ($T_{\text{THz}} = 0.56$, see Eq. (2.16)). In addition, all terahertz power values, displayed in this chapter, are corrected to the attenuation of the employed filter (see Sec. 5.1), which therefore cannot account for this discrepancy either.

Part of the missing terahertz power can be linked to the first parabolic mirror OP1. Figure 5.26a shows a picture of the off-axis mirror from the experimental setup. One can see, that the hole is not exactly in the centre but slightly shifted to the top. Here, the centre is defined by the point of the parabola with a slope of one, diverting a horizontally incoming beam exactly by 90° (see Fig. 5.26b). If the hole is not at this position, the beam cannot leave the mirror perfectly collimated (see Fig. 5.26c). The optimum position

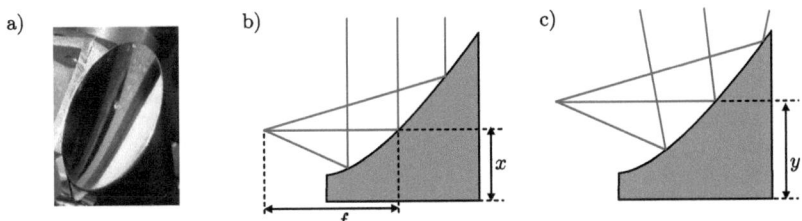

Figure 5.26: *a) Picture of the first off-axis parabolic mirror PM1 with a hole. b) Ideal beam collimation with an off-axis parabolic mirror (side view). The focal length is given by f. The centre of the parabola, where the slope is one, is marked with x. c) The off-axis parabolic mirror is not hit in the centre of the parabola but at the position y. Thus the beam is still diverging when leaving the mirror instead of being well collimated.*

Terahertz wave generation

of the hole would be at a distance x from the bottom, whereas the hole in PM1 is further away at a distance y which might lead to a diverging beam. Our generated terahertz beam already fills the first parabolic mirror OP1 almost entirely (see Fig. 5.2). Therefore, a non-collimated beam will be partly cut off by the second parabolic mirror which may account for power losses.

Another possible explanation for lower output powers could be, that the absorption of the idler wave within the crystal is even higher than 40 cm^{-1}. This value is determined at room temperature [47] while our measurement is performed at 120 °C. Higher temperatures can lead to larger damping, since the vibrating lattice parts are more mobile [93, 94]. In the measurement by Palfalvi et al. the absorption at four different temperatures, 10, 100, 200 and 300 K, is determined. Figure 5.27 shows these values exemplary for a terahertz frequency of 1.5 THz. The behaviour clearly suggests that the absorption for a measurement at almost 400 K, as being the case for our power measurements, should even exceed 50 cm^{-1} (see dashed line in Fig. 5.27). An increase of the absorption from 40 to 50 cm^{-1} would reduce the theoretically predicted output power by almost a factor of two.

Not only the absorption within the crystal but also the absorption of terahertz waves in air is of relevance. The total way through laboratory air from the crystal surface to the Golay cell amounts to almost 20 cm. At a frequency of 1.35 THz, the transmission is high in comparison with 1.4 THz, over 90 % instead of practically 0 % [95] after one meter of propagation. Thus, this cannot be held accountable for a significantly lower terahertz output powers.

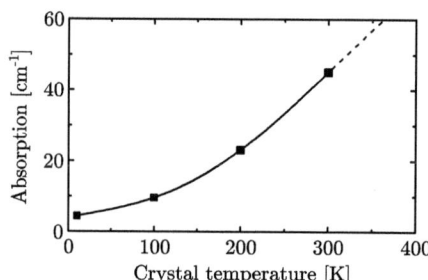

Figure 5.27: *Absorption of MgO-doped lithium niobate as measured by Palfalvi et al. [47]. The squares represent their data points while the solid line acts as a guide to the eye. The dashed line shows a possible extrapolation.*

5.3.2 Properties of terahertz radiation

The wavelength of 222 µm of the terahertz beam and its linear polarisation could be confirmed by the Fabry-Pérot interferometer and polarisation measurements. In the latter measurement, a background of 65 nW detected power remains if the grid is parallel. This probably originates from a slightly imperfect polariser. Although, the dimensions of our wire grids are close to the ones suggested by Goldsmith et al. [86] (see Sec. 5.1.2), they are not ideal or perfectly constructed. The same background phenomenon also occurs in the THz FPI measurement due to the low finesse. Measuring the true linewidth of the terahertz wave remains as a challenge for the continuation of the project.

The beam quality can be quantified by the value of M^2, which compares the divergence of an unknown laser beam with the one of an ideal Gaussian beam [88]. The larger M^2 deviates from $M^2 = 1$, the less perfect is the examined laser beam. To determine the beam quality of our measurement (see Sec. 5.2.4), we assume a wavelength of $\lambda = 222$ µm. The resulting beam focus radius is $w_0 = 1.65$ mm at a distance of 56 mm to the centre of the second parabolic mirror PM2 (Fig. 5.16) [56, 88].

In addition, the analysis gives a beam quality factor of $M^2 = 4.8$. This fit only provides a simple estimate of the beam shape, giving upper limits of its quality. The location of the hole in PM1, transmitting the infrared beams, can be optimised such that the off-axis parabolic mirror is situated ideally inside the cavity with respect to the nonlinear crystal. This should greatly improve the beam quality and enable directing all the terahertz light onto the Golay cell which could result in higher detected powers. The terahertz beam directly at the crystal front facet can intrinsically be a Gaussian beam with $M^2 = 1$.

Due to imperfect collimation, parts of the beam can be cut off by the second parabolic mirror, because the beam can be larger than the mirror diameter. The fit extrapolates a beam radius of 12 mm at the position of PM2, which implies that the terahertz beam fills practically the entire area of the mirror with a diameter of 25 mm. Furthermore, the hole in PM1 might influence the beam quality. A further indication for a non-collimated beam hitting PM2 is the fact, that the location of the focus is 56 mm from the hole position in the parabolic mirror instead of 50 mm.

For more comprehensive examination of the terahertz beam profile, the Golay cell can be covered with a metal foil leaving only a small opening (of some 100 µm). If this Golay cell is then moved through the collimated terahertz beam in a two dimensional scan, one should be able to measure and optimise the intensity distribution of the terahertz beam.

5.3.3 Temperature dependence of the refractive index

Tuning of the generated terahertz waves is achieved by changing the QPM period length as well as the crystal temperature. To compare this tuning behaviour with theoretical predictions, we assume no errors from the infrared contributions to the theory, because, in section 4.1, we have shown that our measured primary OPO process can be well described by the Sellmeier equation of Gayer et al. [46]. Therefore, this equation is taken as a basis for further investigation of the terahertz tuning properties. All deviations between measurement and theory are attributed to the terahertz range in the following analysis.

For the terahertz frequency regime no temperature dependent Sellmeier equation has been available yet. A fit based on the backwards phase matching scheme with the known THz Sellmeier equation [47] in combinations with the infrared ones [46] can be performed. Measurement and prediction only partly agree as shown in Fig. 5.28 exemplary for $\Lambda =$

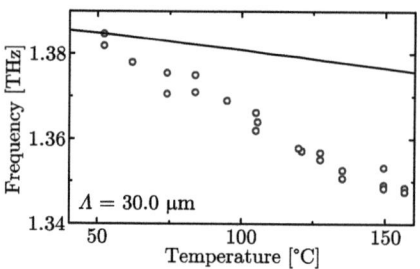

Figure 5.28: *Comparison of measured THz frequencies (blue circles) for a QPM period $\Lambda = 30$ µm and theoretically expected ones (solid line) on the basis of Sellmeier equations [46, 47]. Since the THz Sellmeier equation [47] does not include a temperature dependence, measurement and theory agree less at higher crystal temperatures.*

30.0 µm. Around room temperature (300 K) the theory provides the correct frequency values while at temperatures of around 150 °C the deviation between measurement and theory is 30 GHz.

Based on our data we can deduce a temperature dependence which can be combined with the already existing terahertz Sellmeier equation. Assuming that frequency and temperature dependence decouple, we linearly add a temperature term to the THz Sellmeier equation to match our experimental values:

$$n_{\text{THz}}(\nu) = A + B\tilde{\nu}^2 + C\tilde{\nu}^4 + 0.0013(T - 27), \quad (5.1)$$

with T being the crystal temperature in °C and with $A = 4.94$, $B = 3.7 \times 10^{-5}$ cm^2 and $C = 3 \times 10^{-10}$ cm^4 valid for a crystal temperature of 300 K [47]. Here, $\tilde{\nu}$ is the wave number of the electromagnetic terahertz wave in cm^{-1}. The determined temperature dependence is $dn_{\text{THz}}/dT = 0.0013/$K. These results are graphically illustrated in Fig. 5.29. The refractive index error per temperature made according to this fit is $\pm 0.0001/$K.

To explain the variations of up to 5 GHz between data points taken at the same frequency, one can take a look at the width of the parametric gain for the terahertz process (see Eq. (2.12)). If one neglects absorption for the idler wave, the parametric gain of the cascaded process only has got a FWHM of 1.4 GHz as shown in Fig. 5.30 (black curve). However, we have seen in the experimental sections above, that a frequency jump of

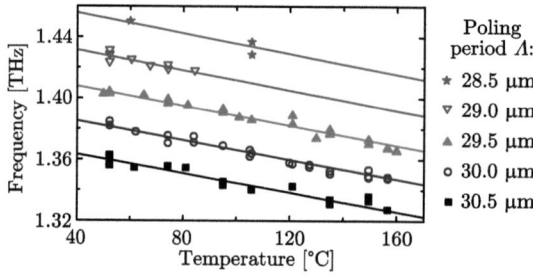

Figure 5.29: *Temperature tuning of the generated terahertz frequency for different poling periods Λ. The symbols represent the measured data points, while the solid lines show the theoretical predictions by a temperature dependent Sellmeier equation (Eq. (5.1)).*

Poling period Λ:
★ 28.5 µm
▽ 29.0 µm
▲ 29.5 µm
○ 30.0 µm
■ 30.5 µm

2.8 GHz can occur in the secondary signal wave while the primary wavelength λ_{s1} remains constant (see Fig. 5.17). Increasing absorption broadens the gain bandwidth. With an idler absorption of $\alpha_{THz} = 40$ cm^{-1} the FWHM is 27 GHz (see Fig. 5.30, red curve). Missing data points at the smaller period lengths might be due to worse coatings for these signal wavelengths and thus less enhancement.

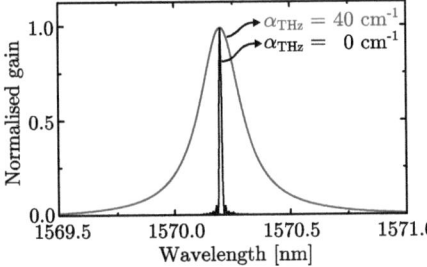

Figure 5.30: *Normalised parametric gain profile of the signal wave from the backwards cascaded parametric process for two different absorptions α_{THz} of the idler wave in the terahertz regime. The FWHM increases from 1.4 to 27 GHz if the absorption grows from 0 to 40 cm^{-1}. Calculations base on Eq. (2.12) with parameters $\lambda_{s1} = 1559$ nm, $L = 2.5$ cm, $\Lambda = 30.0$ µm, $T = 125$ °C.*

Extending the tuning range

The forward terahertz process would be an ideal supplement to the backward process, since it adds a tuning range of 3.1 to 3.6 THz. For the terahertz processes as well as for the infrared OPO, the tuning shown in this work is limited by the crystal structures and mirror coatings available. In general, an additional constraint is present for the terahertz process. Our scheme of terahertz generation bases on a cascaded nonlinear process, making it necessary for both processes, primary and secondary, to be generated within the same crystal structure.

Therefore, currently the detected tuning range has been confirmed with THz power measurements at frequencies between 1.3 to 1.7 THz as generated by the backward process in the QPM periods available. In our ansatz, both, the primary as well as the cascaded parametric process, are phase matched within the same crystal structure. However, this restricts the tuning capabilities. The fundamental limit is reached, once the primary process can no longer be produced by the phase matching structure, as it is the case in our device for QPM periods above 31.4 µm.

Thus, it could be favourable to separate the two processes by using e.g. a dual-structure crystal with two different QPM sections, one for the primary and the other for the cascaded process as exemplified in Fig. 5.31, left side. As shown in the previous chapters, shorter crystals are not necessarily worse, in particular for terahertz generation. The right side of Fig. 5.31 shows the resulting terahertz frequencies from a pump wave at a wavelength of 1550 nm with respect to the poling period. Such a crystal could provide terahertz waves with frequencies ranging from 0.5 to 5 THz out of a single device with poling periods ranging from 10 to 90 µm, only basing on the backwards cascaded process or even larger

Terahertz wave generation

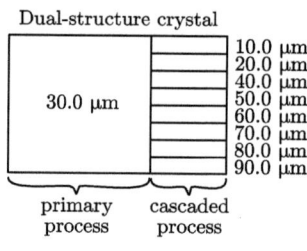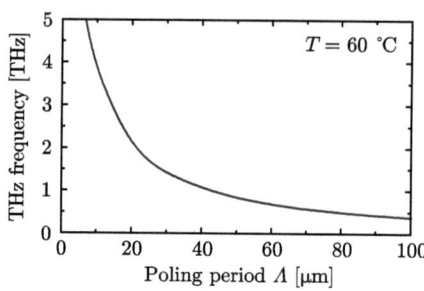

Figure 5.31: Left side: schematic illustration of a dual-structure crystal for terahertz generation. The first section of the crystal can generate the primary process, while the other side is optimised for the cascaded process. Exemplarily the poling periods are chosen to range from 10 to 90 μm. Right side: theoretically expected tuning of the generated terahertz wave from a backwards parametric process with respect to the poling period at a crystal temperature $T = 60\,°C$ and a primary signal wavelength of 1550 nm.

with the forward process in addition. Alternatively two different crystals can be used. In that case, only the crystal for the primary process needs to be put into the pump focus, while the crystal phase matching the cascaded process can be placed into the second signal focus between the two plane mirrors.

A terahertz OPO that is continuously tunable over the entire terahertz range could provide measurements of the refractive indices as well as the nonlinearity at such frequencies. In this way the Sellmeier equations and understanding of nonlinear processes in the terahertz regime can be improved.

5.3.4 Nonlinear coefficient close to a phonon resonance

The nonlinear coefficient, related to the second-order susceptibility, is the driving component for parametric processes but depends on the participating wavelengths. Thus, it is essential to determine its value for frequency conversion into the terahertz regime. In Sec. 2.2.1 it is shown that the pump threshold for parametric oscillation generally depends on this nonlinear coefficient. Therefore, a measurement of the threshold value can be used to deduce the effective value d_{eff} from experimental results, assuming all other contributions in the equation to be known.

We have determined the threshold of the cascaded process for different crystal lengths. From this threshold $P_{th,res}$, the effective nonlinear coefficient for the terahertz range $d_{eff,THz}$ can be deduced via Eq. 2.12. This equation cannot be solved for $d_{eff,THz}$ directly, but one can adjust the value of the nonlinear coefficient until the calculated signal power matches the measured threshold value. Table 5.2, shows the calculated values of $d_{eff,THz}$. This way of measuring the effective nonlinear coefficient in the terahertz range provides a lower

L [cm]	Losses V [%]	$P_{\text{th,res}}$ [W]	$d_{\text{eff,THz}}$ [pm/V]
5.0	1.2	170	109 ± 22
2.0	0.7	330	96 ± 23
1.7	0.7	590	78 ± 25

Table 5.2: *Effective nonlinear coefficient in the terahertz range $d_{\text{eff,THz}}$ determined from the threshold values of the resonant power necessary to start the terahertz parametric process $P_{\text{th,res}}$ for three different crystal lengths L.*

limit of $d_{\text{eff,THz}}$, since the wavemeter needs a minimum incoming power of 1 mW to detect a wavelength component.

The errors that are presented in Tab. 5.2 take uncertainties in the absorption α_{THz}, the beam waist w and the cavity losses V into account. To elaborate the effect of such uncertainties, different values for α_{THz}, w and V are inserted into the parametric gain Eq. (2.12) in combination with Eq. (2.17) as follows: The absorption values in Palfalvi et al. [47] are measured for a maximum temperature of 300 K, which is below our crystal temperatures used. We determine $d_{\text{eff,THz}}$ for $\alpha_{\text{THz}} = (40 \pm 5)$ cm^{-1}. Heating, however, usually increases the absorption, and hence we might underestimate the nonlinear coefficient with this symmetric error of ±5 cm^{-1}. Figure 5.27 suggests that the absorption is even higher a 45 cm^{-1}, which might lead to a further lowering of the measured coefficient, but we have chosen a conservative error estimate to keep it symmetric.

Moreover, there can be an inaccuracy in the cavity losses (mirror reflectivity and residual reflectivity of crystal surfaces) which enters linearly into the threshold equation. Thus, we calculate the influence of a variation in these losses of ±0.1 %. For deducing the electric field from the power threshold of the cascaded process $P_{\text{th,res}}$, we assume the average beam radius of the signal wave to be (100 ± 10) μm. Additional underestimation of $d_{\text{eff,THz}}$ might arise from diffraction effects – only undisturbed plane waves are assumed for the calculation – which could enhance $d_{\text{eff,THz}}$ with respect to our values.

Considering all these issues, our average effective nonlinear coefficient is (94 ± 23) pm/V, with an uncertainty of ±25 %. This value matches well the theoretical predictions by Hebling et al. [96] of $d_{\text{eff,theo}} = 107$ pm/V. Both values are much larger than the effective coefficient for the near infrared, $d_{\text{eff,IR}} = 17$ pm/V [60], which makes nonlinear optics with terahertz waves possible even in the presence of large absorption. Our experimentally determined value for $d_{\text{eff,THz}}$ is also in the same order of magnitude as the predictions of the simple Miller formula (Eq. (2.4)). For the backwards parametric process generating 1.35 THz, the value of Miller's nonlinear coefficient is $d_{\text{Miller,THz}} = 187$ pm/V, leading to an effective coefficient of 119 pm/V. Once more the experimental value is smaller than the theoretical one as expected.

The nonlinear coefficient strongly depends on all interacting waves, as also indicated by Miller's rule (Eq. (2.4)). Therefore, $d_{\text{eff,THz}}$ might be even higher for processes starting from wavelengths, which are closer to phonon resonances themselves, or going to higher THz frequencies. However, this would probably entail higher absorption, which could make such a nonlinear processes technically unrealisable.

With an extended tuning range, one can measure the nonlinear coefficient for processes involving different terahertz frequencies. It is possible that d_{eff} increases even further towards the phonon resonance at 7.6 THz. However, here the absorption will grow as well, providing at some point a limitation of the terahertz tuning. In an exemplary calculation, assuming a constant $d_{\text{eff}} = 100$ pm/V, losses of 0.7 % and a 2.5 cm long crystal, a pump threshold of 1 kW is reached for a terahertz absorption of $\alpha_{\text{THz}} = 650$ cm^{-1}, which would be the case for a frequency of 5 THz. Here, it should be noted that the terahertz frequency enters the threshold equations in such a way, that higher frequencies lead to lower thresholds if all other parameters are kept constant.

5.3.5 Higher-order cascaded processes

In Chapt. 4, it has been shown in Fig. 4.4 that in addition to the secondary parametric process further cascades are possible. The frequency separations of neighbouring peaks is also 1.35 THz as illustrated in Fig. 5.20. Since higher pump powers also provide higher resonant signal powers, the first cascade and thus the following cascades can run more efficiently or set in at all. This scheme of higher-order cascades is illustrated in Fig. 5.32a.

The pump depletion of the primary parametric process drops, when the number of resonant components increases (see Fig. 5.22, right side). The signal wave of the primary process λ_{s1} acts as the pump wave for the secondary process, which is why the primary signal wave is depleted when the cascaded process oscillates. Thus cascaded process is a loss mechanism for the primary OPO process. To analyse this in detail an extended theory is necessary, describing the entire interactions of all parametric processes involved. For this the coupled wave equations would need to constitute of a set of five or even more equations. The power of each signal wave needs to be measured separately to be compared with theory.

Each additional process starts from a different wavelength. Therefore, the generated terahertz frequencies could have slightly altered values due to the dispersion. The change in expected terahertz frequency resulting from the different pump wavelengths is shown in Fig. 5.32b. The terahertz frequency generated by the fourth-order cascade is almost 300 MHz larger than the first terahertz frequency. Here, the primary signal wave is chosen to have a wavelength of 1550 nm.

Yet, within the accuracy of the wavemeter of ±50 MHz per line, all signal waves are equidistant. A difference of 300 MHz in the resulting terahertz frequency cannot be observed. This increased amount of cascades also coincides with higher terahertz output powers (see Fig. 5.23). On the one hand, this can be due to the accumulating terahertz power of each cascaded process. On the other hand, already present THz radiation could seed additional cascaded processes which is illustrated in Fig. 5.33.

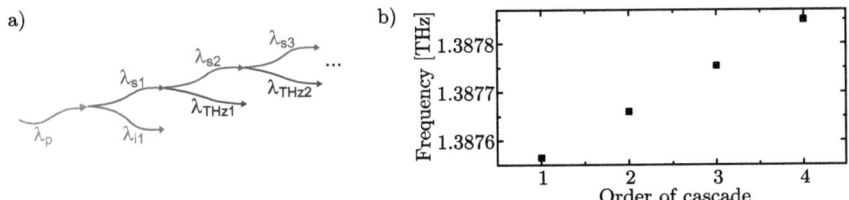

Figure 5.32: a) Scheme of several cascades of parametric terahertz processes. b) Theoretical terahertz frequencies as estimated from the maxima of the gain profiles for the different cascaded processes up to fourth-order with the primary signal wave being at a wavelength of $\lambda_{s1} = 1550$ nm.

Efficient seeding only takes place as long as the parametric gain profile overlaps with the one required by the seed wavelength. For the higher-order cascade this is no problem since the bandwidth of the profile is approximately 30 GHz, while the frequency change of the between first and fourth-order cascade is only 300 MHz (see Fig. 5.32b). Instead of a feeded parametric process one can also interpret the occurrence of higher-order cascades as difference frequency generation between the signal and the terahertz wave.

In Fig. 5.20 one can see that the amplitude of λ_{s3} as detected by the wavemeter is already a factor of three smaller than the one of λ_{s2}. This decrease in amplitude carries on to the last cascade. Since the output power scales with the input powers, the contribution of the higher-order cascades to the total amount of terahertz power can only be a fraction of the THz radiation generated by the first process. This might imply that a high terahertz power enables more cascaded processes rather than the other way around. A way of determining whether the signal waves are equidistant is building a Fabry-Pérot interferometer with a FSR corresponding to an integer multiple of the terahertz wavelength. If all signal waves collapse to a single peak within this FPI, then their spacings should be equal. Of course, they could still be separated by multiples of the FSR but this can be ruled out already with the accuracy of the wavemeter.

If all signal waves were truly equidistant, down to the scale of Hz, they might be able to span a frequency comb [97], provided that the phases between these waves are fixed at the same value. The concept of frequency combs was developed mainly by Hänsch and comprises a tool for very precise frequency measurements [98]. Important ingredients of such a frequency comb are equidistant frequencies which need to span at least an octave, for their absolute positions to be determinable [99]. Those frequency combs are directly produced by femtosecond lasers, but they can also be generated by means of nonlinear

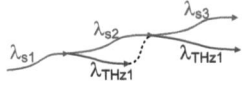

Figure 5.33: The terahertz wave generated in the first cascaded process acts as a seed for the second cascade, which generates the same terahertz wavelength λ_{THz1}.

optics, which has been shown for a third-order nonlinearity [100].

One difficulty in a system like ours is the dispersion of the nonlinear medium, due to which the refractive index changes with wavelength. Therefore, the free spectral range FSR changes with the light frequency inside the cavity, since the length of the light path depends on this refractive index. That is why, a priori the cascaded components cannot be indefinitely equidistant but at some point their frequency separation should differ. However, in our system the change of the FSR over more than ten cascades is less than one MHz which is in the order of the signal linewidth. In Sec. 5.3.5 it is also discussed that efficient seeding is only possible as long as the gain profiles overlap. This limitation might even be reached before the change of FSR comes into play.

Yet, if all terahertz frequencies are generated via seeded parametric processes, they have exactly the same value and their THz powers just add up. In principle, for high conversion efficiencies, it is thus possible to achieve powers larger than the theoretical limit of the Manley-Rowe relations (see Eq. (2.22)): $P_{\text{THz}} = P_{\text{p}}\nu_{\text{THz}}/\nu_{\text{p}}$. These relations only hold for three interacting waves and are therefore no longer limiting for a process involving several cascades. Such an idea has already been reported by Vodopyanov for pulsed terahertz systems [101] and can also be explored in our system.

5.4 Material properties of lithium niobate in the infrared and terahertz frequency regime

In the previous sections, it is stated that the characteristics of MgO-doped lithium niobate in the terahertz regime differ from the ones in the infrared wavelength range. Table 5.3 contrasts important optical features of lithium niobate from these two regions. The refractive index at room temperature for 1 µm, corresponding to 300 THz, is decreased by a factor of two in comparison with the one at 1.4 THz, while the absorption is lower by several orders of magnitude [46, 47].

	n	α [cm^{-1}]	dn/dν [1/THz]	dn/dT [1/K]	d_{eff} [pm/V]
Infrared (300 THz)	2.153	< 0.1	0.0002	0.0004	17
Terahertz (1.4 THz)	5.065	40	0.1793	0.0013	94 ± 23

Table 5.3: *Comparison of material properties of magnesium-doped lithium niobate in the near infrared wavelength range IR (exemplary 300 THz) with the terahertz regime (1.4 THz): refractive index n, the absorption α, the change of n with respect to a frequency interval dν or a temperature interval ΔT, the effective nonlinear coefficient d_{eff}. Data for the infrared is taken from references [46, 60], while the terahertz value for n is from [47] and the other values are determined in this thesis.*

The variation of the refractive index with respect to the frequency difference dn/dν is determined between frequencies of 200 THz (1.5 µm) and 300 THz (1 µm) or 1 and 3 THz, respectively. Here, the variation between IR and THz properties amounts to three orders of magnitude, which shows that close to phonon resonances properties can change dramatically with the frequency. A similar behaviour can be observed, when looking at the modification of n at different temperatures dn/dT. In this case, the refractive index at terahertz frequencies changes three times stronger than the infrared n.

Different nonlinear materials also have varying nonlinear coefficients. An overview of various materials with a non-vanishing second-order nonlinearity d can be found in Shoji et al. [102]. In this article, the highest nonlinear coefficient is given for GaAs, more than five times larger than the one of lithium niobate. However, there the values of d are determined for second harmonic generation in the near infrared, not for parametric processes including terahertz waves. For most materials the nonlinear coefficient for terahertz generation is not known precisely. Considering gallium arsenide, the effective nonlinear coefficient is quantified as only $d_{14,\text{eff}} = 47$ pm/V [103] for a terahertz process. This value is by a factor of two lower than the one we determined for magnesium-doped lithium niobate at 1.35 THz. Yet, the absorption in GaAs is less: 0.6 cm^{-1} [104] instead of 40 cm^{-1} at 1.4 THz. Such a material can also be periodically oriented to achieve phase matching and might be able to provide higher output powers. Its applicability for terahertz generation has already been demonstrated for pulsed systems [103]. Here, one could think of a PPLN

TERAHERTZ WAVE GENERATION

OPO generating radiation with a wavelength of more than 2 µm, which is then sent into a GaAs crystal for terahertz generation. Ideally, this is also performed within a cavity to exploit the intensity enhancement.

Organic materials like DAST can have high nonlinear coefficients for terahertz generation, too, but they cannot survive kilowatts within a resonator [105]. Therefore, one needs to consider issues such as realisable coatings and high intra-cavity power resistance before choosing a certain nonlinear material for terahertz generation.

5.5 Comparison of system performance with that of other methods

In the previous chapters, it has been demonstrated theoretically and experimentally that an optical parametric oscillator is capable of producing continuous-wave terahertz radiation. Its main characteristics are now compared with those of other ways of generating cw terahertz light.

The terahertz frequency range is situated between infrared and microwave frequencies. Therefore, two fundamentally different approaches for terahertz generation can be distinguished (see Sec. 1): starting at higher frequencies, using optical methods, or beginning at lower frequencies and employing electronic technologies. Table 5.4 shows the key properties such as terahertz output power, tuning capability, beam shape or system size of different coherent cw terahertz sources. Some of these methods are already mentioned in the introduction which illustrates advantages and disadvantages of each ansatz. Here, we want to compare their experimental details with the performance of our THz OPO. The output power values in Tab. 5.4 only provide orders of magnitude.

The major issue with systems involving electronic stages is their upper frequency limit. Such a fundamental limit does not exist for nonlinear optics. Besides the achievable absolute frequency values, also the tuning is an important system property for applications such as spectroscopy. Electronic systems and optically pumped lasers are hardly continuously tunable at all. While photomixing and DFG are still constrained by the tunability of the two lasers needed, OPO tuning only depends on the phase matching structures.

It should be emphasised here, that the tuning range mentioned for the THz OPO in Tab. 5.4 is only the one experimentally confirmed within this thesis. In principle much larger tuning ranges, even over the entire terahertz gap could be possible. For example, the forward terahertz process is able to generate up to 3.6 THz emission (see Fig. 4.7). Possible ways for extending the tuning range have already been discussed above and the true boundaries of parametrically generated terahertz radiation still need to be determined. This will be performed in studies following this thesis. Therefore, the THz OPO might be the method of choice for applications requiring large tuning ranges.

If one considers commercialising such an OPO, the current system size of some square

Terahertz wave generation

	BWO [106, 107]	Multipl. [19–21]	Photom. [25, 108]	Opt. laser [26]	QCL [28, 30]	DFG [31, 32]	OPO
Physcial principle	electronic	electronic	optoelectronic	optical	optical	optical	optical
Output power [µW]	10^3	10 at 1 THz	1 at 1 THz	10^5	10^4	10^{-3}	1
Frequency range [THz]	0.1 - 1.5	0.1 - 2	0.1 - 3.5	0.3 - 10	1 - 10	0.5 - 3	1.3 - 1.7
Tuning range [THz]	0.01	0.01	1	none	0.001	0.5 - 3	1.3 - 1.7
Linewidth [MHz]	100	10^{-3}	1	1	0.01	1	1
Beam shape	not Gauß	not Gauß	Gauß	Gauß	not Gauß	Gauß	M^2 < 5
System size [m²]	10^{-2}	10^{-2}	1	few	10^{-2}	few	few

Table 5.4: *Comparison of different methods for continuous-wave terahertz generation. Here, the following abbreviations are used: BWO is a Backward Wave Oscillator, Multipl. are electronic multiplier chains, Photom. is photomixing, Opt. laser means optically pumped laser, QCL stands for Quantum Cascade Laser while DFG represents Difference Frequency Generation via nonlinear optics.*

meters is not very convenient. Additionally, the device requires readjustment and is not a turnkey system so far. However, in contrast to other sources such as quantum cascade lasers, the OPO does not require cooling with liquid helium or nitrogen which is a clear advantage.

Chapter 6
Summary

The terahertz regime (0.1 to 10 THz), which lies between microwaves and infrared wavelengths, is not yet well explored although interactions between molecules produce frequencies in this THz region, which makes this a relevant field not only for fundamental research. However, the generation of terahertz waves is difficult, since electronic systems operate at lower and optical devices at higher frequency levels. Therefore, the term *terahertz gap* formed.

For applications such as high-resolution spectroscopy and telecommunications, terahertz sources should ideally be monochromatic as well as tunable. The scope of this work is the construction of a continuous-wave terahertz emitter based on *nonlinear optics*. Such an ansatz does not have fundamental frequency limitations, only the high absorption of terahertz waves within nonlinear crystals makes THz generation challenging.

Our approach, to solve this problem, is a cascaded parametric scheme: here, the primary signal wave acts as a pump wave for a secondary process, generating a monochromatic terahertz wave. The dispersion in lithium niobate enables both processes to be phase matched within the same periodic crystal structure.

A standard OPO, that operates at infrared wavelengths, forms the basis of our THz system, which is why this device is characterised first. Pumped with light at a wavelength of 1030 nm, its tuning ranges from 1270 to 1840 nm for the signal wave and from 2330 to 5320 nm for the corresponding idler.

Due to the large THz absorption, high infrared powers are needed for THz generation. According to Kreuzer [49], single-frequency operation of continuous-wave OPOs is possible only up to pump powers of 4.6 times the pump threshold P_{th}. This threshold value can be varied by changing crystal length or cavity losses: For our system the best performance is achieved either with a short crystal of 1.7 cm length or an outcoupling mirror of 1.5 % residual transmission in combination with a 5-cm-long crystal. The highest single-frequency powers demonstrated, at the maximum pump power of 18.8 W, are 2.89 W at 3190 nm and 7.63 W at 1520 nm, which are available simultaneously.

This infrared OPO is extended to the first continuous-wave terahertz optical parametric

Summary

oscillator. With a terahertz output power larger than one microwatt it exceeds the power of previous experiments based on nonlinear THz generation by more than two orders of magnitude [32]. Our THz source emits a linearly-polarised, circular beam with a linewidth of several MHz or below. Tuning is presented from 1.3 to 1.7 THz for a terahertz wave traveling antiparallel to its pump wave. A forward process could provide 3.1 to 3.6 THz supplementary.

Our device is employed to investigate the material properties of lithium niobate in the terahertz range. The effective nonlinear coefficient, for a parametric process converting light at a wavelength of 1500 nm to terahertz waves at 1.4 THz, measures (94 ± 23) pm/V. This is thus approximately a factor of five larger than the corresponding coefficient for the near infrared. Refractive index changes with respect to the crystal temperature are also higher in the terahertz region, amounting to $dn_{\mathrm{THz}}/dT = 0.0013$. Higher-order cascaded processes are observed, and their frequency separation and possible extension to a frequency comb is clearly worth further investigations.

The continuous-wave parametric oscillator, that has been presented in this thesis, forms an all-optical terahertz source that does not require cooling. Such a device inhibits high potential for applications in areas like astronomy, telecommunications or high-resolution spectroscopy.

Bibliography

[1] R. Kleiner. *Filling the terahertz gap.* Science **318**, 1254–1255 (2007).

[2] I. Hosako, N. Sekine, M. Patrashin, S. Saito, K. Fukunaga, Y. Kasai, P. Baron, T. Seta, J. Mendrok, S. Ochiai, H. Yasuda. *At the dawn of a new era in terahertz technology.* Proc. IEEE **95**, 1611–1623 (2007).

[3] F. K. Tittel, D. Richter, A. Fried. In *Solid-state mid-infrared laser sources*, ed. I. T. Sorokina, K. L. Vodopyanov, Springer-Verlag, Berlin, 2003.

[4] D. F. Plusquellic, K. Siegrist, E. J. Heilweil, O. Esenturk. *Applications of terahertz spectroscopy in biosystems.* Chem. Phys. Chem. **8**, 2412–2431 (2007).

[5] L. Ho, M. Pepper, P. Taday. *Terahertz Spectroscopy: Signatures and fingerprints.* Nature Photon. **2**, 541–543 (2008).

[6] P. H. Siegel. *Terahertz technology.* IEEE T. Microw. Theory. **50**, 910–928 (2002).

[7] G. Kirchhoff, R. Bunsen. *Chemische Analyse durch Spectralbeobachtungen.* Annalen der Physik und Chemie **110**, 161–189 (1860).

[8] D. Bigourd, A. Cuisset, F. Hindle, S. Matton, E. Fertein, R. Bocquet, G. Mouret. *Detection and quantification of multiple molecular species in mainstream cigarette smoke by continuous-wave terahertz spectroscopy.* Opt. Lett. **31**, 2356–2358 (2006).

[9] M. Tonouchi. *Cutting-edge terahertz technology.* Nature Photon. **1**, 97–105 (2007).

[10] P. H. Siegel. *Terahertz technology in biology and medicine.* IEEE T. Microw. Theory. **52**, 2438–2447 (2004).

[11] M. C. Kemp, P. F. Taday, B. E. Cole, J. A. Cluff, A. J. Fitzgerald, W. R. Tribe. *Security applications of terahertz technology.* Proc. SPIE **5070**, 44–52 (2003).

[12] F. Rutz, M. Koch, S. Khare, M. Moneke, H. Richter, U. Ewert. *Terahertz quality control of polymeric products.* Int. J. Infrared Millimeter Waves **27**, 547–556 (2006).

[13] K. Kawase, Y. Ogawa, H. Minamide, H. Ito. *Terahertz parametric sources and imaging applications.* Semicond. Sci. Technol. **20**, S258–S265 (2005).

[14] P. Mukherjee, B. Gupta. *Terahertz (THz) frequency sources and antennas - A brief review.* Int. J. Infrared Millimeter Waves **29**, 1091–1102 (2008).

[15] J. D. Kraus. *Radio Astronomy.* Cygnus-Quasar Books, 1986.

[16] T. Kleine-Ostmann, K. Pierz, G. Hein, P. Dawson, M. Koch. *Audio signal transmission over THz communication channel using semiconductor modulator.* Electron. Lett. **40**, 124–126 (2004).

[17] M. J. Fitch, R. Osiander. *Terahertz waves for communications and sensing.* Johns Hopkins APL Technical Digest **25**, 348–355 (2005).

[18] S. S. Dhillon, C. Sirtori, J. Alton, S. Barbieri, A. De Rossi, H. E. Beere, D. A. Ritchie. *Terahertz transfer onto a telecom optical carrier.* Nature Photon. **1**, 411–415 (2007).

[19] F. Maiwald, F. Lewen, B. Vowinkel, W. Jabs, D. G. Paveljev, M. Winnewisser, G. Winnewisser. *Planar Schottky diode frequency multiplier for molecular spectroscopy up to 1.3 THz.* IEEE Microwave Guided Wave Lett. **9**, 198–200 (1999).

[20] G. Chattopadhyay, E. Schlecht, J. S. Ward, J. J. Gill, H. H. S. Javadi, F. Maiwald, I. Medhi. *An all-solid-state broad-band frequency multiplier chain at 1500 GHz.* IEEE Trans. Microwave Theory Tech. **52**, 1538–1547 (2004).

[21] S. Schiller, B. Roth, F. Lewen, O. Ricken, M. C. Wiedner. *Ultra-narrow-linewidth continuous-wave THz sources based on multiplier chains.* Appl. Phys. B: Laser Opt. **95**, 55–61 (2009).

[22] E. R. Brown, F. W. Smith, K. A. Mcintosh. *Coherent millimeter-wave generation by heterodyne conversion in low-temperature-grown GaAs photoconductors.* J. Appl. Phys. **73**, 1480–1484 (1993).

[23] E. R. Mueller. *Terahertz radiation sources for imaging and sensing applications.* Photon. Spectra **40**, 60–69 (2006).

[24] K. Sakai. *Terahertz Optoelectronics.* Springer-Verlag Berlin Heidelberg, 2005.

[25] S. Matsuura, M. Tani, K. Sakai. *Generation of coherent terahertz radiation by photomixing in dipole photoconductive antennas.* Appl. Phys. Lett. **70**, 559–561 (1997).

[26] G. Kh. Kitaeva. *Terahertz generation by means of optical lasers.* Laser Phys. Lett. **5**, 559–576 (2008).

[27] R. Kohler, A. Tredicucci, F. Beltram, H. E. Beere, E. H. Linfield, A. G. Davies, D. A. Ritchie, R. C. Iotti, F. Rossi. *Terahertz semiconductor-heterostructure laser.* Nature **417**, 156–159 (2002).

[28] L. Mahler, A. Tredicucci, F. Beltram, C. Walther, J. Faist, B. Witzigmann, H. E. Beere, D. A. Ritchie. *Vertically emitting microdisk lasers.* Nature Photon. **3**, 46–49 (2009).

[29] Y. Chassagneux, R. Colombelli, W. Maineult, S. Barbieri, H. E. Beere, D. A. Ritchie, S. P. Khanna, E. H. Linfield, A. G. Davies. *Electrically pumped photonic-crystal terahertz lasers controlled by boundary conditions.* Nature **457**, 174–178 (2009).

[30] B. S. Williams. *Terahertz quantum-cascade lasers.* Nature Photon. **1**, 517–525 (2007).

[31] J. Nishizawa, T. Tanabe, K. Suto, Y. Watanabe, T. Sasaki, Y. Oyama. *Continuous-wave frequency-tunable terahertz-wave generation from GaP.* IEEE Photon. Technol. Lett. **18**, 2008–2010 (2006).

[32] S. Ragam, T. Tanabe, K. Saito, Y. Oyama, J. Nishizawa. *Enhancement of CW THz wave power under noncollinear phase-matching conditions in difference frequency generation.* J. Lightwave Technol. **27**, 3057–3061 (2009).

[33] M. H. Dunn, M. Ebrahimzadeh. *Parametric generation of tunable light from continuous-wave to femtosecond pulses.* Science **286**, 1513–1517 (1999).

[34] G. W. Baxter, M. A. Payne, B. D. W. Austin, C. A. Halloway, J. G. Haub, Y. He, A. P. Milce, J. Nibler, B. J. Orr. *Spectroscopic diagnostics of chemical processes: applications of tunable optical parametric oscillators.* Appl. Phys. B: Lasers Opt. **71**, 651–663 (2000).

[35] S. E. Harris. *Tunable optical parametric oscillators.* Proc. IEEE **57**, 2096–2113 (1969).

[36] M. Vainio, J. Peltola, S. Persijn, F. J. M. Harren, L. Halonen. *Singly resonant cw OPO with simple wavelength tuning.* Opt. Express **16**, 11141–11146 (2008).

[37] A. Henderson, R. Stafford. *Low threshold, singly-resonant CW OPO pumped by an all-fiber pump source.* Opt. Express **14**, 767–772 (2006).

[38] W. D. Jr. Johnston. *Nonlinear optical coefficients and the raman scattering efficiency of LO and TO phonons in acentric insulating crystals.* Phys. Rev. B **1**, 3494–3503 (1970).

[39] A. Yariv, P. Yeh. *Optical Waves in Crystals.* John Wiley & Sons, New York, 1984.

[40] D. D. Lowenthal. *Cw periodically poled LiNbO$_3$ optical parametric oscillator model with strong idler absorption.* IEEE J. Quantum. Electron. **34**, 1356–1366 (1998).

[41] E Palik. *Handbook of optical constants of solids.* Academic Press, 1985.

[42] R. L. Byer, R. L. Herbst. *Nonlinear infrared generation.* Sringer, New York, 1977.

[43] B. A. E. Saleh, M. C. Teich. *Fundamentals of Photonics.* John Wiley & Sons, Weinheim, 2007.

[44] R. C. Miller. *Optical 2nd harmonic generation in piezoelectric crystals.* Appl. Phys. Lett. **5**, 17 (1964).

[45] C. G. B. Garrett, F. N. H. Robinson. *Miller's phenomenological rule for computing nonlinear susceptibilities.* IEEE J. Quant. Electron. **QE-2**, 328–329 (1966).

[46] O. Gayer, Z. Sacks, E. Galun, A. Arie. *Temperature and wavelength dependent refractive index equations for MgO-doped congruent and stoichiometric LiNbO$_3$.* Appl. Phys. B: Lasers Opt. **91**, 343–348 (2008).

[47] L. Palfalvi, J. Hebling, J. Kuhl, A. Peter, K. Polgar. *Temperature dependence of the absorption and refraction of Mg-doped congruent and stoichiometric LiNbO$_3$ in the THz range.* J. Appl. Phys. **97**, 123505-1 (2005).

[48] R. Fischer. *Abschätzung der Aufbauzeit eines einfachresonanten optischparametrischen Oszillators (Estimation of the rise time of a singly-resonant optical parametric oscillator).* Experimentelle Technik der Physik **22**, 269–271 (1974).

[49] L. B. Kreuzer. *Single and multimode oscillation of the singly resonant optical parametric oscillator.* Proceedings of the Joint Conference on Lasers and Opto-Electronics 52–63 (1969).

[50] W. Brunner, R. Fischer, H. Paul. *Der einfach-resonante optische parametrische Oszillator.* Ann. Phys. **30**, 299–308 (1973).

[51] W. Brunner, H. Paul. In *Progress in optics Vol. XV*, ed. E. Wolf, North-Holland Publishing Company, Amsterdam, 1977.

[52] W. R. Bosenberg, A. Drobshoff, J. I. Alexander, L. E. Myers, R. L. Byer. *93 % pump depletion, 3.5 W continuous-wave, singly resonant optical parametric oscillator.* Opt. Lett. **21**, 1336–1338 (1996).

[53] J. E. Bjorkholm. *Some effects of spatially nonuniform pumping in pulsed optical parametric oscillators.* IEEE J. Quantum. Electron. **QE 7**, 109–118 (1971).

[54] J. M. Manley, H. E. Rowe. *Some general properties of nonlinear elements 1: general energy relations.* Proc. Institute Radio Engineers **44**, 904–913 (1956).

[55] J. M. Manley, H. E. Rowe. *General enery relations in nonlinear reactances.* Proc. Institute Radio Engineers **47**, 2115–2116 (1959).

[56] D. Meschede. *Optik, Licht und Laser.* B. G. Teubner, Stuttgart, 2005.

[57] J. A. Armstrong, N. Bloembergen, J. Ducuing, P. S. Pershan. *Interactions between light waves in a nonlinear dielectric.* Phys. Rev. **127**, 1918–1939 (1962).

[58] L. E. Myers, R. C. Eckardt, M. M. Fejer, R. L. Byer, W. R. Bosenberg, J. W. Pierce. *Quasi-phase-matched optical parametric oscillators in bulk periodically poled $LiNbO_3$.* J. Opt. Soc. Am. B **12**, 2102–2116 (1995).

[59] O. Paul, A. Quosig, T. Bauer, M. Nittmann, J. Bartschke, G. Anstett, J. A. L'Huillier. *Temperature-dependent Sellmeier equation in the MIR for the extraordinary refractive index of 5% MgO doped congruent $LiNbO_3$.* Appl. Phys. B: Lasers Opt. **86**, 111–115 (2007).

[60] J. Seres. *Dispersion of second-order nonlinear optical coefficient.* Appl. Phys. B: Lasers Opt. **73**, 705–709 (2001).

[61] S. E. Harris. *Proposed backward wave oscillation in the infrared.* Appl. Phys. Lett. **9**, 114–116 (1966).

[62] N. E. Yu, C. Jung, C. S. Kee, Y. L. Lee, B. A. Yu, D. K. Ko, J. Lee. *Backward terahertz generation in periodically poled lithium niobate crystal via difference frequency generation.* Jpn. J. Appl. Phys., Part 1 **46**, 1501–1504 (2007).

[63] Y. Sasaki, A. Yuri, K. Kawase, H. Ito. *Terahertz-wave surface-emitted difference frequency generation in slant-stripe-type periodically poled $LiNbO_3$ crystal.* Appl. Phys. Lett. **81**, 3323–3325 (2002).

[64] A. Henderson, R. Stafford. *Intra-cavity power effects in singly resonant cw OPOs.* Appl. Phys. B: Lasers Opt. **85**, 181–184 (2006).

[65] W. R. Bosenberg, A. Drobshoff, J. I. Alexander, L. E. Myers, R. L. Byer. *Continuous-wave singly resonant optical parametric oscillator based on periodically poled $LiNbO_3$.* Opt. Lett. **21**, 713–715 (1996).

[66] A. Ashkin, G. D. Boyd, J. M. Dziedzic, R. G. Smith, A. A. Ballman, J. J. Levinste, K. Nassau. *Optically-induced refractive index inhomogeneities in $LiNbO_3$ and $LiTaO_3$.* Appl. Phys. Lett. **9**, 72–74 (1966).

[67] H. Kogelnik, T. Li. *Laser beams and resonators.* Appl. Opt. **5**, 1550–1567 (1966).

[68] A. E. Siegman. *Lasers.* University Science Books, Mill Valley, California, 1986.

[69] S. Zaske, D. H. Lee, C. Becher. *Green-pumped cw singly resonant optical parametric oscillator based on MgO:PPLN with frequency stabilization to an atomic resonance.* Appl. Phys. B: Laser Opt. **98**, 729–735 (2010).

[70] I. Breunig, J. Kiessling, B. Knabe, R. Sowade, K. Buse. *Hybridly-pumped continuous-wave optical parametric oscillator.* Opt. Express **16**, 5662–5666 (2008).

[71] R. W. Boyd. *Nonlinear Optics.* Academic Press, San Diego, 2003.

[72] M. M. J. W. van Herpen, A. K. Y. Ngai, S. E. Bisson, J. H. P. Hackstein, E. J. Woltering, F. J. M. Harren. *Optical parametric oscillator-based photoacoustic detection of CO_2 at 4.23 µm allows real-time monitoring of the respiration of small insects.* Appl. Phys. B: Lasers Opt. **82**, 665–669 (2006).

[73] H. Su, Y.-Q. Qin, H.-C. Guo, S.-H. Tang. *Periodically poled $LiNbO_3$: Optical parametric oscillation at wavelengths larger than 4.0 µm with strong idler absorption by focused Gaussian beam.* J. Appl. Phys. **97**, 113105 (2005).

[74] M. M. J. W. van Herpen, S. Li, S. E. Bisson, S. T. Hekkert, F. J. M. Harren. *Tuning and stability of a continuous-wave mid-infrared high-power single resonant optical parametric oscillator.* Appl. Phys. B: Lasers Opt. **75**, 329–333 (2002).

[75] S. E. Harris. *Threshold of multimode parametric oscillators.* IEEE J. Quantum Electron. **QE 2**, 701–703 (1966).

[76] D. A. Long. *Raman Spectroscopy.* McGraw-Hill, New York, 1977.

[77] A. S. Barker, R. Loudon. *Dielectric properties and optical phonons in $LiNbO_3$.* Phys. Rev. **158**, 433–445 (1967).

[78] S. Kojima. *Composition variation of optical phonon damping in lithium niobate crystals.* Jpn. J. Appl. Phys. **32**, 4373–4376 (1993).

[79] Yu. Serebryakov, N. Sidorov, M. Palatnikov, V. Lebold, Ye. Savchenkov, V. Kalinnikov. *The influence of Mg^{2+}, Gd^{3+} and Ta^{5+} admixtures on cation structural ordering in lithium niobate single crystals.* Ferroelectrics **167**, 181–189 (1995).

[80] U. T. Schwarz, M. Maier. *Frequency dependence of phonon polariton damping in lithium niobate.* Phys. Rev. B **53**, 5074–5077 (1996).

[81] R. Quispe-Siccha, E. V. Mejía-Uriarte, M. Villagrán-Muniz, D. Jaque, J. García Solé, F. Jaque, R. Y. Sato-Berrú, E. Camarillo, J. Hernández, H. Murrieta. *The effect of Nd and Mg doping on the micro-Raman spectra of $LiNbO_3$ single-crystals.* J. Phys.: Condens. Matter **21**, 145401 (2009).

[82] A. V. Okishev, J. D. Zuegel. *Intracavity-pumped Raman laser action in a mid-IR, continuous-wave (cw) MgO:PPLN optical parametric oscillator.* Opt. Express **14**, 12169–12173 (2006).

[83] A. Henderson, R. Stafford. *Spectral broadening and stimulated Raman conversion in a continuous-wave optical parametric oscillator.* Opt. Lett. **32**, 1281–1283 (2007).

[84] M. J. E. Golay. *Theoretical consideration in heat and infra-red detection, with particular reference to the pneumatic detector.* Rev. Sci. Instrum. **18**, 347–356 (1947).

[85] H. Kogelnik. *Imaging of optical modes - resonators with internal lenses.* Bell Syst. Tech. J. **44**, 455–494 (1965).

[86] Paul F. Goldsmith. *Quasioptical Systems.* IEEE Press, New York, 1998.

[87] A. E. Costley, K. H. Hursey, G. F. Neill, J. M. Ward. *Free-standing fine-wire grids: Their manufacture, performance, and use at millimeter and submillimeter wavelengths.* J. Opt. Soc. Am. **67**, 979–981 (1977).

[88] M. W. Sasnett. *Propagation of multimode laser beams - The M^2 factor.* In *The physics and technology of laser resonators*, ed. D. R. Hall, P. E. Jackson, Hilger, London, 1989.

[89] A. Dobroiu, C. Otani, K. Kawase. *Terahertz-wave sources and imaging applications.* Meas. Sci. Technol. **17**, R161–R174 (2006).

[90] B. Ferguson, X. C. Zhang. *Materials for terahertz science and technology.* Nature Mat. **1**, 26–33 (2002).

[91] Y. C. Shen, T. Lo, P. F. Taday, B. E. Cole, W. R. Tribe, M. C. Kemp. *Detection and identification of explosives using terahertz pulsed spectroscopic imaging.* Appl. Phys. Lett. **86** (2005).

[92] L. Ho, R. Mueller, K. C. Gordon, P. Kleinebudde, M. Pepper, T. Rades, Y. Shen, P. F. Taday, J. A. Zeitler. *Monitoring the film coating unit operation and predicting drug dissolution using terahertz pulsed imaging.* J. Pharm. Sci. **98**, 4866–4876 (2009).

[93] I. P. Ipatova, A .A. Maradudi, R. F. Wallis. *Temperature dependence of width of fundamental lattice-vibration absorption peak in ionic crystals: 2. Approximate numerical results.* Phys. Rev. **155**, 882–895 (1967).

[94] I. P. Ipatova, A. V. Subashie, A. A. Maradudi. *Temperature dependence of fundamental lattice vibration absorption by localized modes.* Ann. Phys. **53**, 376–418 (1969).

[95] A. Danylov. *THz laboratory measurements of atmospheric absorption between 6 % and 52 % relative humidity.* Submillimeter-Wave Technology Laboratory, University of Massachusetts, Lowell, 2006.

[96] J. Hebling, A. G. Stepanov, G. Almaasi, B. Bartal, J. Kuhl. *Tunable THz pulse generation by optical rectification of ultrashort laser pulses with tilted pulse fronts.* Appl. Phys. B: Lasers Opt. **78**, 593–599 (2004).

[97] T. Udem, R. Holzwarth, T. W. Hänsch. *Optical frequency metrology.* Nature **416**, 233–237 (2002).

[98] J. Reichert, R. Holzwarth, T. Udem, T. W. Hänsch. *Measuring the frequency of light with mode-locked lasers.* Opt. Commun. **172**, 59–68 (1999).

[99] T. Udem, J. Reichert, T. W. Hänsch, M. Kourogi. *Accuracy of optical frequency comb generators and optical frequency interval divider chains.* Opt. Lett. **23**, 1387–1389 (1998).

[100] P. DelHaye, A. Schliesser, O. Arcizet, T. Wilken, R. Holzwarth, T. J. Kippenberg. *Optical frequency comb generation from a monolithic microresonator.* Nature **450**, 1214–1218 (2007).

[101] K. L. Vodopyanov. *Optical generation of narrow-band terahertz packets in periodically inverted electro-optic crystals: conversion efficiency and optimal laser pulse format.* Opt. Express **14**, 2263–2276 (2006).

[102] I. Shoji, T. Kondo, A. Kitamoto, M. Shirane, R. Ito. *Absolute scale of second-order nonlinear-optical coefficients.* J. Opt. Soc. Am. B **14**, 2268–2294 (1997).

[103] K. L. Vodopyanov, M. M. Fejer, X. Yu, J. S. Harris, Y. S. Lee, W. C. Hurlbut, V. G. Kozlov, D. Bliss, C. Lynch. *Terahertz-wave generation in quasi-phase-matched GaAs.* Appl. Phys. Lett. **89**, 141119 (2006).

[104] K. L. Vodopyanov. *Optical THz-wave generation with periodically-inverted GaAs.* Laser Photon. Rev. **2**, 11–25 (2008).

[105] M. Jazbinsek, L. Mutter, P. Gunter. *Photonic applications with the organic nonlinear optical crystal DAST.* IEEE J. Quantum Electron. **14**, 1298–1311 (2008).

[106] G. Kantorowicz, P. Palluel. *Backward Wave Oscillators.* In *Infrared and Millimeter Waves*, ed. K. Button, Academic Press, 1979.

[107] B. Gorshunov, A. Volkov, I. Spektor, A. Prokhorov, A. Mukhin, M. Dressel, S. Uchida, A. Loidl. *Terahertz BWO-spectrosopy.* Int. J. Infrared Millimeter Waves **26**, 1217–1240 (2005).

[108] S. Matsuura, H. Ito. In *Terahertz Optoelectronics*, ed. K. Sakei, Springer-Verlag, Berlin, 2005.

List of publications with own contributions

- I. Breunig, R. Sowade and K. Buse, "Limitations of the tunability of dual-crystal optical parametric oscillators," Opt. Lett. **32**, 1450 (2007).

- I. Breunig, M. Falk, B. Knabe, R. Sowade, K. Buse, P. Rabiei and D. Jundt, "Second harmonic generation of 2.6 W green light with thermo-electrically oxidized undoped congruent lithium niobate crystals below 100 °C," Appl. Phys. Lett. **91**, 221110 (2007).

- I. Breunig, J. Kießling, B. Knabe, R. Sowade and K. Buse, "Hybridly-pumped continuous-wave optical parametric oscillator," Opt. Express **14**, 5662 (2008).

- I. Breunig, J. Kießling, R. Sowade, B. Knabe and K. Buse, "Generation of tunable continuous-wave terahertz radiation by photomixing the signal waves of a dual-crystal optical parametric oscillator," New J. Phys. **10**, 073003 (2008).

- J. Kiessling, R. Sowade, I. Breunig, K. Buse and V. Dierolf, "Cascaded optical parametric oscillations generating tunable terahertz waves in periodically poled lithium niobate crystals," Opt. Express **17**, 87 (2009).

- R. Sowade, I. Breunig, J. Kiessling and K. Buse, "Influence of pump threshold on single frequency idler output of singly-resonant optical parametric oscillators," Appl. Phys. B **96**, 25 (2009).

- R. Sowade, I. Breunig, I. Cámara-Mayorga, J. Kiessling, C. Tulea, V. Dierolf and K. Buse, "Continuous-wave optical parametric terahertz source," Opt. Express **17**, 22310 (2009).

- R. Sowade, I. Breunig, C. Tulea and K. Buse, "Nonlinear coefficient and temperature dependence of the refractive index of lithium niobate in the terahertz regime," Appl. Phys. B **99**, 63 (2010).

Acknowledgements

During the time of conducting my thesis a lot of people helped in many different ways. Here, I would like to take the opportunity of saying thank you to some of them:

Prof. Dr. Karsten Buse for giving me the chance to be part of his group and work on this excellent project. He always had trust in my abilities and encouraged me to go abroad and apply for scholarships. In addition, he provided a great working environment at the Heinrich Hertz foundation chair of the Deutsche Telekom AG.

Prof. Dr. Stephan Schlemmer for conducting a review of my thesis and for the friendly collaborations with his group at Cologne university, which stimulated further improvement of the OPO. Many thanks also to Prof. Dr. Hans Werner Hammer and Prof. Dr. Peter Vöhringer for taking over the secondary reviews.

The OPO team and primarily its head Dr. Ingo Breunig. His scrutiny and rigorous leadership pushed everyone in the team forward and he was always approachable for advice and practical aid. My diploma students, Jens Kießling and Cristian Tulea, for extending this work with their contributions. In particular, we all greatly benefitted from Jens Kießling, who will now continue the terahertz OPO project, by his ability to construct small devices such as the *Höllenmaschine*.

I thank the entire Hertz group! In particular, Dr. Akos Hoffmann who solves almost any technical and electronic challenge. Prof. Dr. Boris Sturman who visited our group regularly and for him handling coupled wave equations seemed like the easiest thing to do. Raja Bernard, for helping with any problems not involving physics.

Support of the Deutsche Telekom Stiftung is gratefully acknowledged, especially for making my stay in Stanford possible. Additional sponsorship came from the Bonn Cologne Graduate School of Physics and Astronomy.

Some people have been part of my life for much longer than just my thesis years: My cousin Eva who was always there for me and without her I might not have started physics to begin with. Sonja and Bettina who became my friends during school time. Bettina, although the exchange to the USA took place long ago, we're still standing here shoulder to shoulder.

My boyfriend Ingo for cheering me up and enriching my life. I am deeply thankful for his support, serenity and understanding. He successfully convinced me that the two of us are unbeatable.

Last but not least, my family who was by my side the entire way. Without the financial help of my parents, my studies would have been impossible. However, far more important is their emotional support, to which also my sister Ramona immensely contributed.

Thanks a lot!

I want morebooks!

Buy your books fast and straightforward online - at one of world's fastest growing online book stores! Environmentally sound due to Print-on-Demand technologies.

Buy your books online at
www.morebooks.shop

Kaufen Sie Ihre Bücher schnell und unkompliziert online – auf einer der am schnellsten wachsenden Buchhandelsplattformen weltweit! Dank Print-On-Demand umwelt- und ressourcenschonend produziert.

Bücher schneller online kaufen
www.morebooks.shop

KS OmniScriptum Publishing
Brivibas gatve 197
LV-1039 Riga, Latvia
Telefax +371 686 204 55

info@omniscriptum.com
www.omniscriptum.com

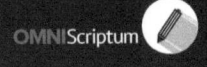

Printed by Books on Demand GmbH, Norderstedt / Germany